SERIES PREFACE

Food and food production have never had a higher profile, with food-related issues featuring in newspapers or on TV and radio almost every day. At the same time, educational opportunities related to food have never been greater. Food technology is taught in schools, as a subject in its own right, and there is a variety of food-related courses in colleges and universities - from food science and technology through nutrition and dietetics to catering and hospitality management.

Despite this attention, there is widespread misunderstanding of food - about what it is, about where it comes from, about how it is produced, and about its role in our lives. One reason for this, perhaps, is that the consumer has become distanced from the food production system as it has become much more sophisticated in response to the developing market for choice and convenience. Whilst other initiatives are addressing the issue of consumer awareness, feedback from the food industry itself and from the educational sector has highlighted the need for short focused overviews of specific aspects of food science and technology with an emphasis on industrial relevance.

The *Key Topics in Food Science and Technology* series of short books therefore sets out to describe some fundamentals of food and food production and, in addressing a specific topic, each issue emphasises the principles and illustrates their application through industrial examples. Although aimed primarily at food industry recruits and trainees, the series will also be of interest to those interested in a career in the food industry, food science and technology students, food technology teachers, trainee enforcement officers and established personnel within industry seeking a broad overview of particular topics.

Leighton Jones
Series Editor

Campden & Chorleywood Food
Research Association Group

Key Topics in Food
Science and Technology – No. 10

Chemical analysis of foods: an introduction

Leighton Jones

Campden & Chorleywood Food Research Association Group comprises
Campden & Chorleywood Food Research Association
and its subsidiary companies
CCFRA Technology Ltd   CCFRA Group Services Ltd   Campden & Chorleywood Magyarország

© CCFRA 2005

Campden & Chorleywood Food
Research Association Group

Chipping Campden, Gloucestershire, GL55 6LD UK
Tel: +44 (0) 1386 842000   Fax: +44 (0) 1386 842100
www.campden.co.uk

Information emanating from this company is given after the exercise of all reasonable care and skill in its compilation, preparation and issue, but is provided without liability in its application and use.

The information contained in this publication must not be reproduced without permission from the CCFRA Publications Manager.

Legislation changes frequently. It is essential to confirm that legislation cited in this publication and current at the time of printing is still in force before acting upon it. Any mention of specific products, companies or trademarks is for illustrative purposes only and does not imply endorsement by CCFRA.

valid analytical measurement

VAM Helpdesk, LGC, Queens Road, Teddington, Middlesex, TW11 0LY
Tel: +44(0)20 8943 7393   email: vam@lgc.co.uk   www.vam.org.uk

© CCFRA 2005
ISBN: 0 905942 72 8
A catalogue record for this book is available from the British Library.

PREFACE TO THIS VOLUME

Big decisions are made on the basis of food analysis. Is the food safe? Is it what it says on the label? Are the declared ingredients, nutritional information and other declarations accurate? Does it conform to legal limits for pesticide and veterinary residues? Does the ingredient or product conform to specifications agreed between the retailer, manufacturer and/or supplier? Often, these and many other questions can only be answered through chemical analysis. And if this is not done properly, the answer is likely to be wrong. This, in turn, means that tonnes of product might unnecessarily be re-worked or disposed of, wasting considerable amounts of material and energy. Worse still, unsafe or fraudulent product might reach the marketplace. So to be of use, the analysis has to be right: it has to be done properly.

Proper chemical analysis of food is, however, easier said than done. Finding and measuring the amount of an individual chemical in a complex mixture of thousands of others is sometimes like looking for a needle in a haystack. It requires exacting science and a diligent, organised approach to every aspect of the analysis - from sourcing, selecting and preparing the samples right through to interpreting the result. And between these ends of the 'analytical chain', a lot can go wrong.

Fortunately guidance is available. The Valid Analytical Measurement (VAM) programme funded by the UK Department of Trade and Industry was created to help organisations carry out analytical measurements competently and accurately. It also helps users of analytical services to understand why best practice is important in terms of the reliability of the results they get and on which they base their decisions. The VAM programme advocates six seemingly obvious but very important principles that encapsulate best practice. This book, produced as part of the VAM programme, describes the chemical analysis of food by looking at the main methods used in the context of their day-to-day application within industry, enforcement and government. In so doing, it describes the principles of VAM and illustrates their importance. It emphasises that no matter how sophisticated the instrument or skilled the analyst, if best practice is not adopted the analysis is no more than a waste of time, effort and money. As with all the books in this series the intention is to be illustrative rather than comprehensive, and to provide leads for the reader to follow the subject in more detail.

Leighton Jones, CCFRA

ACKNOWLEDGEMENTS

The preparation of this book was supported under contract with the Department of Trade and Industry as part of the National Measurement System Valid Analytical Measurement (VAM) programme. The VAM programme aims to improve the quality of analytical measurements in the UK. Aspects of the VAM programme relating to chemical measurement are managed by LGC.

I am grateful to the following who contributed information or reviewed the manuscript, in whole or part, at various stages in its preparation:

LGC	Vicki Barwick and Peter Farnell
CCFRA	Nick Byrd, Howard Davies, Megan Davies, Dr. John Dooley, Paul Drake, Dr. Mike Edwards, Louise Gearey, Rob Levermore, Brian McLean, Christina Oscroft, Sue Salmon, Margaret Voyiagis and Dr. Steven Walker
VAM Food Reference Material User Group	Simon Freeman (Nestlé), John Ruegg (Weetabix Ltd), Ian Jerrum (Hampshire Scientific Services) and Alan Bruce and Jane Binks (Glasgow Scientific Services)

Some of the examples used to illustrate specific points are drawn from other CCFRA publications and I am grateful to the authors of these (as cited). Finally, particular thanks are due to Janette Stewart (CCFRA) for the book's design and artwork.

NOTE

All definitions, legislation and codes of practice mentioned in this publication are included for the purposes of illustration only and relate to UK practice unless otherwise stated. Specific products, companies or methods are also mentioned at various points, again for illustrative purposes, and this does not imply endorsement by CCFRA or the VAM programme.

CONTENTS

1.	Introduction	1
1.1	Why analyse food?	4
1.2	Purpose defines approach	5
1.3	The 'analytical chain'	7
1.4	Where things can go wrong	7
1.5	Some rules for getting it right	9
2.	The right approach, the right result	11
2.1	Purpose of the analysis	13
2.2	Sampling and samples	18
2.3	Method development: from R&D to trials	22
2.4	Validation and validated methods	26
2.5	Quality controls and standardisation	29
2.6	Quantitative measurements	31
2.7	Measurement uncertainty, its sources and its estimation	33
2.8	Accreditation	36
2.9	Proficiency testing schemes	37
2.10	Choosing a laboratory to do an analysis	39
3.	Techniques in chemical analysis	42
3.1	Sample preparation and analyte extraction	43
3.2	Gravimetric methods	47
3.3	Chromatography	50
	3.3.1 TLC	53
	3.3.2 HPLC	54
	3.3.3 GLC/GC	55
	3.3.4 Detection systems	58
3.4	Mass spectrometry	60
3.5	Atomic emission and absorption spectrophotometry	64
3.6	Colorimetry/spectrophotometry	66
3.7	X-ray microanalysis	69
3.8	Titration	71

3.9	Dumas and Kjeldahl	72
3.10	pH	73
3.11	Water activity	75
3.12	Immunoassay	75
3.13	Electrophoresis	77
3.14	DNA methods	80

4.	Uses and examples of food analysis	83
4.1	Compositional analysis	84
4.2	Authenticity	89
4.3	Detecting contaminants	94
4.4	Pesticide and veterinary residue detection	99
4.5	Surveillance exercises	103
4.6	Legislation	107
4.7	Flavour, off-flavour and taint analysis	109
4.8	Checking suitability for purpose	112
4.9	Product development	114
4.10	Analysis as a research tool	115

5.	Complications and compromises in food analysis	118
5.1	Dietary fibre: an empirical approach	119
5.2	Global migration: an empirical approach	120
5.3	From nitrogen to protein and meat content	121
5.4	Energy content	126
5.5	Fat content	128
5.6	Sodium versus salt	129
5.7	Vitamins are not one group	131

6.	Conclusions	133
7.	Glossary and common acronyms	135
8.	Selected references, further reading and websites	140

About CCFRA and VAM	144

1. INTRODUCTION

Food is composed of many hundreds of different chemicals. Some of these are an essential part of our diet - providing the raw materials and energy necessary for the growth, development, maintenance and day-to-day activities of the human body. Other components may be of no nutritional value but nevertheless make the food pleasant and appealing, contributing to its colour, flavour and texture for example, and defining the food's quality and essential characteristics. Yet others are added to enhance these qualities or to delay spoilage. There are also chemicals which contribute to neither the food's nutritional value nor its character; they just happen to be there by virtue of having been part of the plant or animal from which the food has been derived.

Table 1 lists, as broad groups, the major and minor components of food. When looking at this it is worth remembering that whilst we think of protein, for example, as one food component, foods can contain perhaps thousands of different proteins. Likewise foods can contain many different fats and sugars. And the starches from different crops (e.g. potato, maize, wheat), whilst all called starch, are chemically slightly different. In other words, most foods are complex mixtures of chemicals, so that finding any one chemical in a food is very much like looking for a needle in a haystack.

Chemical analysis involves using one or more of a wide range of techniques to find the 'needle' in question - i.e. to find one chemical or group of chemicals in a complex mixture. The range of analyses of analytes in foods is tremendous - from minute amounts of pesticides or flavour compounds to components that make up nearly all of the product (e.g. fat in butter, refined sugar). The exact method used depends on the nature of the chemical being sought (the analyte), the type of food (commonly called the matrix), and the purpose of the analysis. For example, completely different approaches would be taken for assessing the protein content of wheat (an important feature in assessing its suitability for breadmaking), identifying the fats in a vegetable oil (e.g. to check its authenticity) or checking for pesticide residues on fruit (e.g. for compliance with legal limits).

Table 1 - The major and minor components of food

Major	Minor
Water	Vitamins
Proteins	Minerals
Carbohydrates - such as sugars and starch	Flavour and taste compounds
Fibre (undigestable material such as complex polysaccharides)	Additives
Lipids (fats and oils)	

As almost all food is derived from other livings things (plants, animals and microbes), it naturally contains thousands of biochemicals. Also, during preparation and processing these often react together to generate many more. Roasting a piece of meat, for example, will link together sugars and proteins (part of the so-called Maillard reaction) to form hundreds of compounds not present in the raw ingredients and which give the cooked product its unique characteristics. For Further information on food composition see Hutton (2002).

Some methods are designed simply to identify whether or not a chemical is present (qualitative methods) whereas others can be used to determine how much of the chemical is present (quantitative). If the method is quantitative, the amount of analyte can be expressed in various ways (see Box 1).

At this point it is perhaps worth briefly distinguishing between chemical analysis (i.e. using analysis to look for particular chemicals) and physical analysis (e.g. analysing features such as viscosity, colour or texture). Confusion can arise because physical properties such as these result from specific chemical components in the food. For example, the pigments present and their chemical state will influence product colour, while the presence of different forms of starch will influence the product's viscosity. Both the pigments and the starch can be measured chemically, and the properties they confer on the food can be measured physically; the former type of analysis falls within the scope of this book but the latter does not.

Box 1 - Expressing amounts

There are several ways of expressing the amount of an analyte in a food. The most common of these is to express the mass of the analyte in a known amount of the food - that is the concentration of the analyte in the food. For example, the amount of fat in a sample of nuts might be expressed as grams of fat per 100g of nuts (g/100g) or grams of fat per kilogram of nuts (written as g/kg or g kg^{-1}). In many cases the amount of analyte is many times smaller than the amount of the food and so is expressed as milligrams per kilogram (mg/kg or mg kg^{-1}), micrograms per kilogram (µg/kg or µg kg^{-1}), or even nanograms or picograms per kilogram (ng/kg and pg/kg). Using the unit of grams, the following table illustrates the relationship between these amounts:

Number of grams	Prefix, unit and symbol
1000	kilo - kilogram - kg
0.001	milli - milligram - mg
0.000 001	micro - microgram - µg
0.000 000 001	nano - nanogram - ng
0.000 000 000 001	pico - picogram - pg

The same principle applies to volumes and, for a drink for example, the amount of analyte might be expressed as weight per volume (litre) as g/l or mg/l for example.

The table illustrates that a milligram is a millionth of a kilogram. For this reason, mg/kg are sometimes expressed as parts per million (ppm). Likewise, there are a billion micrograms in a kilogram so that µg/kg are sometimes expressed as parts per billion (ppb), a trillion nanograms in a kilogram so that ng/kg are sometimes expressed as parts per trillion (ppt), and a quadrillion picograms in a kilogram so that a pg/kg is a part per quadrillion (ppq). These terms are also used for weights in volumes - so, for example, 1 mg/l is also referred to as one part per million. However, although still quite widely used, terms such as ppm, ppb and ppt are generally discouraged as they can cause confusion - not least because the words billion and trillion are non-scientific with different meanings in the UK and US.

continued....

Another way of expressing amounts is in percentage terms, where 1 g in 100 g (or 10 g in 1 kg) is referred to as 1%. This is often qualified as 1% w/w (weight by weight) and, on a similar basis 1 g in 100 ml can be expressed as 1% w/v (weight by volume) and 1 g in 1 litre as 0.1% w/v.

The following table, based on a VAM poster, shows how these different terms inter-relate, and gives them some context by illustrating the volume a single sugar cube (6 g) would need to be dissolved in to achieve that amount:

One sugar cube in a:	Equivalent amounts		
Teapot (0.6 l)	10g/l	10g/kg	1%
Bucket (6 l)	1g/l	1g/kg	0.1%
Tanker lorry (6000 l)	1mg/l	1mg/kg	1ppm
Tanker ship (6 million l)	1µg/l	1µg/kg	1ppb
Reservoir (6 billion l)	1ng/l	1ng/kg	1ppt
Bay (6 trillion l)	1pg/l	1pg/kg	1ppq

Source: VAM poster - How little is "little"? See *www.vam.org.uk*

1.1 Why analyse food?

Foods are subjected to chemical analysis for many reasons including, for example:

- Determining composition - perhaps to assess compliance with company specifications, legal standards or labelling declarations. This includes verifying the authenticity of ingredients or products, or assessing products for payment purposes (e.g. paying farmers for supplying milk to processors).
- Checking for contaminants - for example to ensure that pesticide or veterinary residues do not exceed legal limits, to look for allergen (e.g. nut material) in allergen-free material, or to check the form of a contaminant (e.g. organic mercury is more toxic than free mercury).
- Verifying authenticity and detecting adulteration - that is, making sure that a product conforms to information presented on the label or in other associated documentation.

- Assessing suitability for purpose - for example, checking that a product designed for coeliacs (people with gluten intolerance) is free of gluten, or testing wheat for its suitability for breadmaking.
- Supporting product development - for example, by verifying the consistency of composition of formulations of particular ingredients, such as the composition of vegetable oils in a low-fat spread.
- Analysis of flavour volatiles (e.g. looking at changes in flavour of materials produced or stored under particular conditions) or other components in foods (e.g. colours in sweets).
- Surveillance exercises - often carried out by government agencies and sometimes by companies as a way of monitoring the safety, quality and legal compliance of specific foods or products. This also includes amassing data on intake of nutrients or potential contaminants by consumers or specific groups of consumers.
- Calibrating instruments for rapid analysis of food samples (perhaps on-line or near-line).
- Research - for example assessing the factors which influence the formation of desirable compounds (e.g. colour complexes or volatile profiles) or undesirable compounds (e.g. acrylamide, natural toxicants) in particular products or processes and how these can be controlled.

These overlapping uses of food analysis are covered in greater detail in Chapter 4, where they are illustrated through many more examples. However, just from the above short list, the importance of analysis in assurance and policing of food safety, quality and authenticity is obvious. Moreover, the importance of conducting analyses properly should be equally obvious: erroneous analytical data could compromise food safety, undermine product quality and lead to incorrect labelling information. This could result in legal action against a company, severely damaging its reputation and jeopardising its commercial success and brand image.

1.2 Purpose defines approach

The exact approach taken in an analysis will depend on various factors, including the likely level of the analyte, the matrix, the degree of precision required and so on.

However, it should also depend on the purpose of the analysis, as different approaches can yield different results. It is therefore important to be clear about how the results are going to be used before planning and doing (or commissioning) the analysis. For example, there are many methods available for analysing foods for pesticide residues. Some of these can be used to detect many pesticides at the same time (and so are called multi-residue screens) whereas others detect just one or a few closely related pesticides (called targeted analysis).

Another example is provided by analysis of aflatoxins (a group of fungal toxins). In some cases, a simple qualitative analysis (i.e. presence / absence above a certain threshold) might be adequate. On other occasions a full quantitative analysis might be appropriate as this provides information on how much of each type of aflatoxin is present. A third example is provided by analysis of foods for salt - that is, sodium chloride. This is usually determined by analysing the amount of sodium and using this to calculate the salt content. However, in instances where a food contains sodium from other sources (e.g. from the raising agent sodium bicarbonate or the flavour enhancer monosodium glutamate), calculation via chloride might give a more accurate estimate.

In some instances, the analyte itself is not a specific chemical, and so has to be defined in terms of the analysis. For example, dietary fibre is the indigestible material in food. Taking this broadly, some define dietary fibre as a mixture of complex polysaccharides (including pectins, gums, mucilages and cellulose) as well as non-polysaccharide material such as lignin and undigested protein and lipid. Others argue that the definition should be limited to include plant polysaccharides other than starch (so called non-starch polysaccharide NSP) together with lignin. And there are further variations on these definitions. Without a clear chemical definition of dietary fibre, there is no single method of analysis acceptable to everyone. For example, the AOAC (Association of Official Analytical Chemists) method involves weighing the material after extraction and enzymic digestion. In contrast, the Englyst method is more complicated, involving measurement of individual sugars following breakdown of the polysaccharides. What this means in practice is that the term dietary fibre is taken to mean the quantity present in the food measured by the chosen (and named) analytical technique. So it is important to use the appropriate method (especially where comparisons are being drawn between foods). This is discussed further in Section 5.1.

Further examples of how 'purpose defines approach' are outlined in Section 2.1 and will be evident in many of the examples of uses of food analysis described in Chapters 4 and 5.

1.3 The 'analytical chain'

It is easy to think of analysis in terms of the analytical test alone. The test itself, however, is just one in a series of steps, all of which are important if the test is to generate meaningful data. Figure 1 maps out the typical stages involved in conducting an analysis. All this pre-supposes that the test method has been properly validated and is subject to on-going quality assurance and control, and that the laboratory checks its performance through participation in proficiency testing schemes - points which are all discussed later.

Overall, the analysis is conducted to generate information, and this information will lead to action. If things go wrong at any one step in this chain, then the wrong action might be taken with serious consequences.

1.4 Where things can go wrong

If something goes wrong at any point in the analytical chain it can, and usually will, affect the result. Although the analysis itself is often thought of as the most important part of the sequence of events, this is not necessarily the case. For example, if the wrong question is posed so that the purpose of the analysis is not clear (i.e. there is not a clear specification), the wrong technique might be used and the results could be useless. Similarly, the analysis would be of no use if the sample was not representative of the batch it was chosen to represent or if it had been stored under conditions that affect the analyte. If, during storage or sample preparation, the sample became contaminated with analyte from the adjacent sample or a standard solution (this is 'cross-contamination'), the result would be meaningless. And the report would be useless if it contained the results for the wrong samples.

Figure 1 - The main steps in the analytical chain

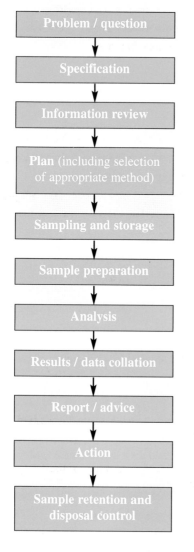

All parts of the chain are important. For example, the reason the analysis is being conducted is because there is a specific problem that needs addressing or question that needs answering. By being clear on this it is possible to create a specification of what is required of the analysis and to consider this in the light of other information to hand. The whole process can then be planned so that the appropriate samples can be obtained and handled, stored and prepared properly.

After analysis, the results will need to be calculated, checked and interpreted so that a report can be produced - stating the results in relation to the question originally posed. On the basis of the report, advice can be issued and decisions taken that result in some action. Samples will usually be retained for some time after analysis in case re-analysis is necessary; this often involves a formal system for sample disposal (e.g. written consent from client).

This might sound protracted, and experienced analysts will do much of it automatically, but breaking down the process like this illustrates that meaningful analysis is about more than just doing the test.

So whilst there are well recognised systems for validating analytical methods and checking that they work as expected (see Chapter 2), quality management systems are also important for ensuring good practice along the whole analytical chain. Guaranteeing that mistakes will not happen is not possible, but minimising the opportunities for mistakes through adoption of good practice is an extremely important part of quality assurance and good laboratory practice.

1.5 Some rules for getting it right

Good practice can go a long way to assuring the reliability of test results. This can be achieved through the consistent application of some basic principles at all points in the analytical chain. In the UK, the VAM (Valid Analytical Measurement) programme was established to promote best practice in chemical analysis - not just of foods but across all sectors where chemical analysis plays a role. Table 2 lists

Table 2 - The six principles of VAM

Principle 1	Analytical measurements should be made to satisfy an agreed requirement
Principle 2	Analytical measurements should be made using methods and equipment which have been tested to ensure they are fit for purpose
Principle 3	Staff making analytical measurements should be both qualified and competent to undertake the task
Principle 4	There should be a regular independent assessment of the technical performance of a laboratory
Principle 5	Analytical measurements made in one location should be consistent with those elsewhere
Principle 6	Organisations making analytical measurements should have well defined quality control and quality assurance procedures

the six principles advocated as part of the VAM programme to help assure reliable data. These are discussed in more detail in subsequent chapters, where the importance of good analysis is illustrated through numerous examples. The fundamental message, however, is simple: making the right decision requires having the right result, which in turn depends on using the right approach.

2. THE RIGHT APPROACH, THE RIGHT RESULT

It is important for a laboratory to take the right approach if it is to get the right result. The analytical chain, described in Section 1.3, illustrates that there are many points at which errors can arise that could compromise the reliability of the analysis. For example, the samples might be stored or handled inappropriately so that the levels of the analyte change, or results might be inaccurate because equipment is not properly calibrated. This can lead to decisions which waste time, money, materials and effort or have even more serious consequences (e.g. compromise product safety, incur legal action). This chapter looks in more detail at some of the main factors that can influence analyses. It also describes some of the systems that can be used to prevent problems occurring and assure the reliability of results.

An alternative to thinking of analysis as a chain of events, is to consider it in terms of inputs and outputs. The output will be the results, usually embodied in a report that might also carry some interpretation of the results and a discussion of their significance. There are many inputs into the analysis, as depicted in Figure 2. Examples include the equipment used and its calibration, laboratory staff, methods and chemicals, and general facilities. In addition to these 'direct' inputs, aspects like method validation, quality control routines, laboratory management and participation in proficiency schemes, will also all influence the analysis and result. All of these are important.

Laboratories can take many practical measures to ensure that the 'inputs' are managed properly, so making the outputs reliable. In effect, these practical measures are just ways of implementing the principles of VAM described earlier (see Section 1.5). For example, calibrating equipment and validating methods help to ensure that the methods and equipment are fit for purpose (VAM Principle 2). Using appropriately qualified staff, with on-going training programmes, helps ensure they are competent to undertake the analysis (VAM Principle 3). Participation in proficiency testing schemes enables labs to assess their own performance (Principle 4). The laboratory

Chemical analysis of foods

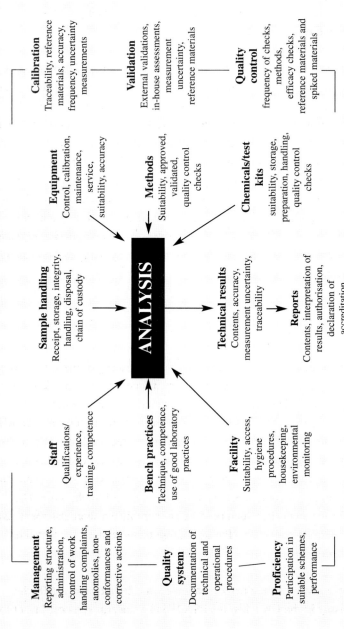

Figure 2 - Analysis in terms of inputs, outputs and laboratory management

Analysis can be thought of in terms of inputs and outputs. The sample is processed by staff using appropriate facilities, equipment and methods to generate results that go into a report. More general laboratory procedures and activities will also influence the quality of the end result - including management of personnel and resources, method validation, and use of quality control procedures. Figure adapted from The Campden Laboratory Accreditation Scheme (CCFRA, 2003).

can then also use these systems to demonstrate to clients that they are competent to undertake the required analyses (e.g. by achieving accreditation). This gives the client confidence in the laboratory and is good for the laboratory's business as it helps to attract and retain clients.

The following sections explain and provide examples of some of these procedures, controls and systems. They also illustrate that, while the systems help assure the reliability of the results and demonstrate their competence to clients, they also carry a cost - in terms of materials (e.g. chemicals or test kits to perform the analysis) and considerable time and effort. The laboratory has to recover this either through the price of the analysis or, for an in-house laboratory of a food company, through an increase in its running costs. Although this can act as a disincentive for a company to chose a laboratory that undertakes these activities, the additional cost of analysis is likely to be negligible compared to the costs of using a laboratory that generates incorrect results leading to decisions that themselves become expensive mistakes. This is discussed further in Section 2.10 and reinforced by many of the examples covered in Chapter 4.

2.1 Purpose of the analysis

The importance of being clear about the purpose of the analysis has already been raised (see Section 1.2). One of the examples mentioned includes pesticide analysis - specifically, multi-residue screening versus targeted analysis of individual compounds. A company may want to know the level of say, chlormequat on wheat, to check conformance with legal limits (maximum residue levels or MRLs). For this, it would need to get targeted analysis carried out, as this pesticide is not detected in a multiresidue screen. The results of a multi-residue screen could not be used, therefore, to check compliance with the relevant MRL in this case. The specification for the analysis, which need be no more than a short statement, should make the exact requirement clear. 'Please analyse these samples for pesticides' would not suffice.

Another example given was the determination of salt. In September 2004 this became a big issue in the UK when the FSA (Food Standards Agency) launched a major campaign to reduce salt (and hence sodium) intake. This was based on evidence linking excessive sodium intake to increased blood pressure (hypertension) and increased risk of developing cardiovascular disease. Salt is sodium chloride. In solution it dissociates into its two components - sodium ions and chloride ions. It is not possible to analyse for sodium chloride as such, but it is possible to analyse for sodium (by atomic absorption spectrophotometry for example - see Section 3.5) and chloride (by titration for example - see Section 3.8). Knowing the amount of sodium or chloride then allows calculation of the amount of sodium chloride (salt) as each gram of salt contains approximately 0.39 g sodium and 0.61 g chloride. So, multiplying the sodium content by 2.54 will give the equivalent salt content or multiplying the chloride content by 1.65 will give the equivalent salt content.

But there are pitfalls. For example, foods often contain other sources of sodium. For example sodium bicarbonate is commonly used as a raising agent, and sodium nitrite may be used to cure meat. In foods containing these, basing the salt estimate on sodium content will overestimate the salt levels. Some foods contain chloride from non-salt sources (e.g. from potassium chloride) and in this case the level of salt calculated from chloride content could be overestimated. So the choice of whether to base the analysis on sodium or chloride is important: "Analyse the sample for salt content" is not an adequate specification.

Furthermore, under the Food Labelling Regulations 1996 there is a requirement to declare sodium (not salt) levels on the product label if a claim is made in relation to salt (e.g. low salt). Under current UK Food Advisory Committee guidance, a 'low salt' claim can be made if the food contains less than 40 mg sodium per 100 g of food (or 100 ml of drink). But 'salt' means more to many consumers than does the term sodium, and the FSA campaign focused on targets for salt intake (e.g. 6 g per day as an upper limit for an adult). So some manufacturers supplement the sodium declaration with a figure for 'salt equivalent' (i.e. sodium x 2.54). This example illustrates the need for the person requesting the analysis and the analyst to be clear about the analyte (salt, sodium or chloride), to be aware of possible complicating factors arising from the food (non-salt sodium or chloride), and to consider the use of the information.

A third example is the chemical assessment of rancidity. This is a sensory phenomenon, where a breakdown in fats within a food can lead to the formation of off-flavours. Chemical tests can be used to measure the level of off-flavour compounds or their precursors and so can give a good indication of the degree of rancidity or the likelihood of rancidity occurring (see Box 2). However, as rancidity occurs in different ways, the results of any one test have to be interpreted with caution. The type of test used often depends on the foodstuff and the likely stage of rancidity. If the wrong test is selected it is possible to get a result which suggests the product is not rancid when sensory analysis suggests it is or vice versa. Again, being clear about the purpose of the test (e.g. to measure rancidity or the potential for rancidity) is important in determining the analytical approach taken.

So, it is important for those commissioning analyses to be clear about the reasons for the analysis and how the result is to be used. Discussing these with the analyst will enable them to verify that the purpose is consistent with the approach and to highlight potential pitfalls.

Box 2 - Chemical analysis and rancidity

Rancidity is a sensory problem arising from oxidation of fats to form off-flavours. Whilst rancid food is not, in itself, unsafe, it can be extremely unpleasant. For some sectors (e.g. fats and oils, meat products) rancidity is an important quality issue. Chemical analysis of rancidity is quite difficult and, in addition to the usual analytical skill of the chemist, it requires a lot of experience and knowledge. This is because there are various analytical approaches available and the one chosen depends to a large extent on the food type and the problem being addressed.

The figure overleaf outlines the breakdown process of fats to off-flavour molecules. Initially, triglycerides are broken down (hydrolysed) by the enzyme lipase to release free fatty acids (FFA). This 'hydrolytic rancidity' can cause problems in itself - especially in products such as palm or coconut oil which

continued....

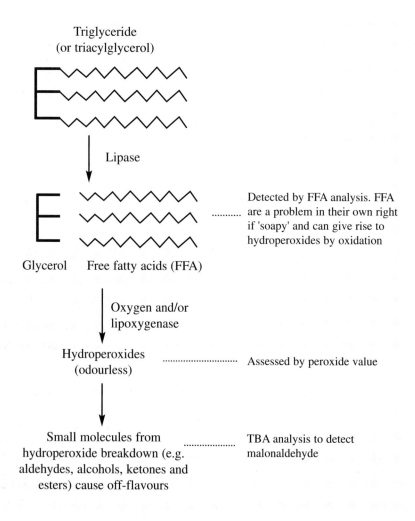

contain short fatty acids such as lauric acid. These can impart a soapy flavour, which is why hydrolytic rancidity is sometimes called soapy rancidity. Even in products where the FFAs do not themselves cause off-flavours, the problem does not end there. They can undergo oxidation (oxidative rancidity - stimulated by the presence of oxygen, lipoxygenase enzymes or light) to form hydroperoxides. Although these are odourless and do not contribute to off-flavours, they tend to be unstable and can break down into a wide range of smaller molecules including aldehydes, ketones, alcohols and esters, which do contribute to off-flavours.

continued....

Three approaches illustrate the importance of using the most appropriate methods - free fatty acid (FFA) analysis, peroxide value (PV) and thiobarbituric acid (TBA). FFA analysis can give a useful indication of rancidity in oils (e.g. coconut) where FFAs themselves cause problems. It is also used extensively as a general indication of the condition and edibility of pure oils and fats, and can give an indication of the likelihood of oxidative rancidity occurring. However, in some cases, where oxidative rancidity has occurred, the FFA level might be quite low - and so of little use on is own.

PV analysis gives an indication of the level of peroxides in a fat sample and can help in an assessment of the progress of oxidative rancidity. A high peroxide value can indicate the potential for further oxidation and the formation of off-flavour molecules. Again, however, care has to be taken: as peroxides are generally short-lived, PV values can be quite low in foods in which rancidity is advanced. This is particularly the case with peroxides formed from unsaturated fatty acids (e.g. from vegetable oils) as they tend to be more reactive than those from saturated fats (e.g. from meat products). PV analysis is therefore more suited to analysis of saturated fats. TBA is a reagent that reacts with malonaldehyde to form a red compound that can be quantified spectrophotometrically (see Section 3.6) to generate a TBA number - which increases as rancidity develops.

The TBA test can be applied directly to the food sample without fat extraction. Both PV and FFA are carried out on pure fat samples - a fat extraction step is required if they are to be applied to food. The methods can be used individually or in combination, and often require 'baseline' measurements of fresh material for comparison. The analyst needs to carefully assess the options and requirements of the analysis to ensure that the method (or methods) used are those most relevant to the information sought. Otherwise much time and effort can be expended for very little gain and without solving the product problems.

References:

Allen, J.C. and Hamilton, R.J. (1994) Rancidity in foods. Third edition. Blackie Academic and Professional. ISBN 0-7514-0219-2

Levermore, R. (2004) Rancidity in fresh and stored pork products. Meat International **14**(7) 16-18

2.2 Sampling and samples

There are various aspects to sampling and sample handling. These include, for example, the decision on which products to test, the statistical aspects of getting a representative sample, and the practical aspects of physically taking, handling, storing and preparing samples. With the physical act of taking the sample, there is also the distinction between the sampling that goes on 'at source' (which is often beyond the control of the analyst) and the 'sub-sampling' that occurs in the laboratory to obtain the final test portion for analysis. All of these aspects are extremely important because if the sample that is analysed is not appropriate, then the rest of the analysis will be a waste of time.

The decision on which of a very wide range of products should be analysed will usually be risk-based - whether it is for routine quality control or monitoring purposes by a company or for surveillance by an enforcement body. Certain products are more at risk of quality defects, contamination or adulteration, for example, and will be targeted more often than those that are less at risk. Many examples of this are given in Chapter 4, but Box 4 illustrates why chilli-containing products were targeted for monitoring during a scare about contamination with Sudan I. To give another example, CCFRA has published a tool to help companies identify which products should be prioritised for pesticide analysis - on the basis of risk - as part of a product monitoring system (Stanley and Knight, 2001). Prioritising what to analyse allows most effective use of resources and clear 'purpose' for the analysis.

The statistics of sampling is beyond the scope of this book, but a few general points are worth emphasising (for more on this see Schutz, 1984 and Jewell, 2001). Most chemical analyses are destructive - that is, they involve grinding up the sample and extracting the analyte so that the part of the product that is analysed is lost. Clearly, 100% testing (i.e. analysis of all of the product) is not only impractical (due to the sheer volumes produced) but also not desirable (as the manufacturer would have no product left to sell). It is therefore necessary to take samples, and these are usually quite small compared to the total production volume.

Although small, such samples should nevertheless be representative of the batch or portion of the batch being assessed. This means that they should be randomly selected: the sampler should have access to all of the 'units' that make up the production run (or relevant part of the production run), each unit should be identifiable, and each should have an equal chance of being selected. Also, to be representative of the entire lot, small sample units should be randomly drawn from many locations, rather than the entire sample from one randomly selected location. For example, any temptation for the sampler simply to take samples from the top carton or edge of the pallet should be resisted. There are exceptions to this. For example, the analysis might be targeted at the first product off a production run to check if it is contaminated with product from the previous run. For the purposes of seeking 'defective units' within a production run - in part or in its entirety - there is a simple rule of thumb which also helps put the idea of sampling into a practical context (see Box 3).

Another consideration is homogeneity - that is, whether the analyte is evenly distributed through the product (and therefore equally likely to be detected in any sample). This is more likely, and easier to achieve, with certain products than with others. For example, a drum of vegetable oil that has been thoroughly mixed could be assumed to be reasonably homogenous, unlike a sack of peanuts in which only one peanut might be contaminated with aflatoxin but to a high level (a so-called 'hotspot'). A range of guidance documents is available to help companies, enforcement authorities and analysts with the statistical aspects of sampling including, for example, Food Standards Agency (2004), Crosby (1996) and various BSI publications.

Once a representative sample has been taken, the sample has to be handled, stored and prepared properly so that the sample is not altered in any way that will affect the analysis. Considerable attention is paid to the conditions under which the sample is stored and handled, as some analytes are susceptible to certain factors (e.g. temperature, light) which can influence their levels and therefore affect the analysis. Some examples are given in Table 3.

Box 3 - How many samples?

One question that often arises is 'How many samples should be taken for analysis?'. For a product in discrete units (e.g. jars, burgers, biscuits, apples) there is a simple rule of thumb to link the number of analyses conducted to the likelihood of there being a problem. Imagine, for example, that a fresh produce distributor wants to check that none of the apples in a consignment contains a particular pesticide. How many apples should be analysed to see if any contain the pesticide?

To be 100% certain that none of the apples contain the pesticide it would be necessary to analyse every apple. And as the test is destructive, this would mean destroying the whole batch - which defeats the purpose of sampling and analysis.

To be 95% confident that no more than '1 in n' apples contain the pesticide it would be necessary to analyse 3n randomly selected apples and find no pesticide. So, for example, to be 95% confident that no more than 10% (1 in 10) of the apples contain the pesticide, it would be necessary to analyse 30 (i.e. 3 x 10) randomly selected apples and find none containing pesticide. Similarly to be 95% confident that no more than 5% (1 in 20) of the apples contain the pesticide, it would be necessary to analyse 60 (i.e. 3 x 20) randomly selected apples and find none containing the pesticide.

Looking at this more generally, to be 95% sure of finding 'defectives' when the proportion of defectives is 1 in n, it is necessary to analyse 3n randomly selected units:

To detect a level of:	It is necessary to test:
1 in 10 defectives	30 units
1 in 100 defectives	300 units
1 in 500 defectives	1,500 units
1 in 1,000 defectives	3,000 units
1 in 10,000 defectives	30,000 units

This approach can be applied to an entire production run or to a particular part of a run if appropriate – for example to target a particular batch code or product from a particular part of the supply chain. Note that the sample size required depends on the proportion of defectives and not on the number of items in the batch. So, coming back to the apple example, the table above applies irrespective of whether the consignment contained 100 apples or 100,000 apples.

However, it is not just about storage: the entire logistics of sample handling is very important. Having one individual or team of individuals with the designated responsibility of sample care can save the analyst valuable time, improve the efficiency of sample throughput and reduce the scope for errors (see Table 4). The team can oversee and standardise the chain of custody: from sample collection and/or receipt, logging, storage (including conditions used and space allocation) and sample flow, to client authorisation for disposal and then disposal itself (Hughes *et al.* 2004). This can lead to significant economy of scale.

Table 3 - Correct storage of samples

Analyte / analysis	Recommended storage
Aflatoxins	Frozen and in darkness (UV light degrades aflatoxins)
Alcohol content of spirits	Refrigerated with airtight lid and no headspace (prevents evaporation)
Carbohydrates (e.g. starch, sugar)	Frozen but moisture-proof (prevents water absorption)
Dithiocarbamate (pesticide)	Immediate analysis (breaks down if frozen and defrosted)
DNA tests	Frozen
Quinine	In darkness (broken down by light)

The table gives examples of recommended storage conditions for samples, on the basis of the susceptibility of the analyte to degradation or other loss (e.g. evaporation). In many instances the nature of the matrix will also need to be considered (e.g. powders such as coffee and milk powder will absorb moisture). In other cases, the analysis might be invalidated if the sample is stored for any length of time, irrespective of the conditions. For example, once rancidity in meat has been initiated it will continue even in frozen samples. Further information and practical guidance on storage of food samples for chemical analysis can be found in CCFRA (2002).

Table 4 - Activities of the sample handling team that can free the analyst to concentrate on the analysis

- Efficient collection of samples, if necessary, for example from retail outlets for surveillance work
- Gathering and recording information on the sample and the history of the food from which it came, especially where these have a bearing on the analysis (e.g. whether the sample has been frozen)
- Checking condition (e.g. for damage) and appropriateness (e.g. quantity) of samples immediately on receipt
- Obtaining additional or replacement samples if required
- Photography of samples if required
- Systematic logging of samples and maintenance of records (e.g. entry on to LIMS - Laboratory Information Management System)
- Efficient and appropriate use and maintenance (e.g. temperature monitoring) of storage areas
- Appropriate sample preparation just before analysis (e.g. cleaning, peeling, cooking, blending)
- Obtaining client authorisation for and organisation of sample disposal

2.3 Method development: from R&D to trials

The requirements for food analysis change as new issues come along - whether it is new ingredients, new contaminants or new processes. A couple of examples illustrate this. The food scare centring on Sudan I required the adaptation and validation of a method to detect this adulterant in chilli powder and products in which chilli powder is used as an ingredient. Because Sudan I is not a permitted food additive, there was no existing method for analysis of foods for Sudan I. In this case the analyte was known but there was no specific method for its detection. An existing method for other food colours was adapted and validated so that it could be used to detect the illegal use of this dye in the necessary products (see Box 4).

A second example is provided by food irradiation. The use of irradiation for preservation of a limited range of products (e.g. herbs and spices, poultry) was

Box 4 - Sudan dyes

The Sudan dyes first became an issue for some sectors of the food industry in 2003. Sudan I was discovered in chilli powder that had been imported from India and used in the preparation of a range of products (see table). The incident provides a good example of how food adulteration and safety can be closely linked. For unscrupulous traders, the motivation behind adding Sudan dye to the chilli powder was to enhance the powder's redness but, as Sudan I had been found to cause cancer in laboratory animals, and it had not been approved for food use, it's presence in food should be avoided.

Types of products known to be affected with Sudan dyes

Chilli powders	Relishes, chutneys and pickles
Spice mixes containing chilli	Vegetable oil
Chilli meals	Seasonings
Various curries and curry sauces	Cook-in sauces
Samosas and similar spicy snacks	

Sudan I is used for industrial colouring - for example, in solvents, oils, waxes, and shoe and floor polishes. The only colourings permitted in food are those specified in the UK Colours in Food Regulations (1995). Sudan I is not specified in these regulations and so is not permitted in food. Because these dyes are not permitted food additives there was, at the time of the first incident, no generally accepted, validated method for their detection and quantification in food. Revisions to a method for food colourings, and validation using a range of appropriate food matrices (e.g. chilli powder, sauces, ready-meals) containing known amounts of the dye, enabled deployment of a method that was fit for purpose. Following a major incident with Sudan I in February 2005, the method was used by enforcement authorities and industry in support of ingredient traceability records as part of a major product withdrawal.

References:

Food Standards Agency website *www.food.gov.uk* - provides additional background on Sudan dyes in food

Jones, L. and Davies, H. (2005) Seeing red - VAM in the food industry. VAM Bulletin No. 32 pp13-16

legally permitted in the UK for the first time in 1991 (see Leadley *et al.* 2003 for more on the preservation aspects). Its use was subject to strict controls including licensing of premises, documentation of processes, labelling of products, and auditing. However, to be fully effective, these controls and the surveillance schemes necessary to police these controls, rely on having analytical methods that can distinguish between irradiated and non-irradiated food. In this case, there was no known analyte. Consequently, during the late 1980s and through the 1990s, various research projects were undertaken including those funded by the UK and other European Governments to explore a range of analytical approaches. All involved looking at irradiation-induced changes in the food - first to find a 'marker' for the process (e.g. a chemical present in irradiated but not in non-irradiated food) and then to develop a reliable method for the routine detection of this marker. A variety of methods emerged. One example is use of gas chromatography with mass spectrometry (GC-MS - see Section 3.4) to detect 2-alkylcyclobutanone. This compound forms on irradiation of fatty acids and can be used as a marker for irradiation treatment of poultry and some seafoods.

Developing a new method to the point where it can be widely adopted and routinely used for reliable food analysis can take a long time. In the case of detecting irradiation treatment of food, for example, it was several years before methods had been developed, refined, validated and trialled to the point where they could be used in surveillance exercises. This is because there are various stages that a method must pass through in its development before analysts can really be sure it is performing adequately and delivers sufficiently reliable results. These stages are outlined in Figure 3.

Method development should always start with a specific question, namely 'What is the problem that needs addressing?' Answering this question needs careful consideration. For example, is the analyte known (as in the Sudan case above) or not (as in the irradiation case)? Is it sufficient simply to know whether an analyte is present or is it important to know the level of the analyte? Which foods will need to be tested? And so on. Once the problem is clearly defined, potential approaches can be considered and preliminary investigations can begin. It might be that an existing method can be adapted or that the approach is at least obvious. In developing methods for the detection of GM material, developing a DNA test for the modified

Figure 3 - The method development chain

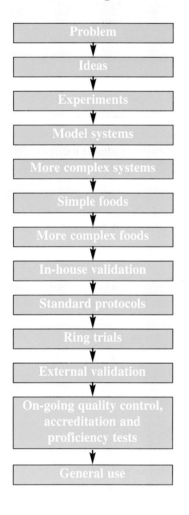

Methods are usually developed in response to a specific problem. Initial ideas will be refined and lead to experiments, often using simple model systems and gradually more complex systems until the method can be validated within the laboratory (e.g. using materials with known amounts of analyte added - including reference materials) and then tested across several laboratories (ring trial). If successful, the method might then be validated as part of a formal validation scheme (see Section 2.4), become included within the scope of a laboratory's accreditation (see Section 2.8) and be used in proficiency tests (see Section 2.9).

Not all methods will follow this entire chain or necessarily follow all these steps in this order, but reputable analysts will want to ensure that a method is performing sufficiently well before bringing it into service.

gene is an obvious approach because the gene provides a good marker for the material and is likely to be detectable in both unprocessed and processed materials.

It is likely that methods will initially be explored using very simple foods (or non-food model systems) before they are used on more realistic samples. Standard materials - that is, 'positives' which are known to contain the analyte and 'negatives' known not to contain the analyte - will be important in validation studies by providing controls. These will include analysing a wide range of materials containing known amounts of the analyte (including reference materials) to generate data on method performance (see Section 2.4).

Protocols might be developed and refined on an iterative basis - by finding variations that improve detection or reduce interference, for example. Good sample preparation is likely to be important, especially if the method has to be applied to a wide range of food materials and can only work on 'clean' extracts (i.e. those from which many impurities have been removed). The method might then undergo a simple inter-laboratory trial (ring trial) - the objective of which is to test the method rather than the laboratory (proficiency tests do the latter - see Section 2.9). Eventually the method might be independently validated by some third party (see Section 2.4). Not all methods in routine use pass through all of these stages, but enough work should be done on the method to ensure it is reliable and fit for purpose.

2.4 Validation and validated methods

For an analyst to be confident that a method does the job that is required - i.e. that it measures what needs to be measured - the method should be validated. This basically means the method should be tested to make sure it works properly and to give the analyst an idea of how well it performs - at least for the materials and analytes to which it will be applied. The latter point is important: if a method works well with certain foods but not with others, this should not prohibit its use where it works well. Perhaps, more commonly, it is the case that the same analytical method can be applied but that different sample preparation systems are required for different food types - for example, for pesticide residue screening different extraction procedures are often used for fatty, aqueous and dried foods.

So what does method validation involve? In general terms, any laboratory that sets out to use a particular method should check that the method works in their hands - even if it is an established and widely used method - and generate documented evidence that it does so. This is called 'in-house' validation because the laboratory is concerned with checking that the method performs adequately in the context of its own systems, facilities and personnel (i.e. in-house). A guide published as part of the VAM programme (see LGC/VAM 2003b) lays out a practical approach to help laboratories to validate methods they have previously not used. Some of the issues that need to be addressed in method selection and validation are listed in Table 5.

The exact approach depends on whether the method is completely new, based on minor changes to an existing method, or is a verification of the performance of a previously validated method. However, it generally includes, amongst other things, assessing whether the method detects only the analyte of interest and whether it performs reliably from occasion-to-occasion, operator-to-operator and, perhaps, laboratory to laboratory. It will also involve looking at the extent to which the result is influenced by minor variations in procedures, and the importance of documenting activities in which these features are explored. Some of these are covered in later sections (cross-referenced from Table 5) and/or in the glossary, as well as in much more detail and with practical guidance in LGC/VAM (2003b).

A method can also be subject to third party validation - where it is assessed by independent laboratories, perhaps as part of a formal method validation scheme. For example, AOACI (Association of Official Analytical Chemists International) runs several schemes for third party validation. The AOAC Performance Tested Methods[SM] Program is designed to provide independent third party review of proprietary methods such as test kits. This provides manufacturers with independent assessment of their method in relation to their performance claims. Such independent assessment is useful for both the manufacturer when marketing the test and for the laboratory using the test kit. Even more rigorous is the AOAC Official Methods[SM] Program, which typically involves scrutiny of in-house method validation data, a statistical and safety review, and assessment by at least 8 independent laboratories - and typically takes at least 12 months.

Table 5 - Examples of issues requiring attention during method selection and validation

Issue	Comment
Property (analyte) to be measured	Being clear about what is being analysed and the reason for the analysis is an important part of deriving a specification for the analysis (see Section Section 1.2)
Matrix	To which foods / drinks will the method be applied? Ensure that validation work is on representative types of material as this can greatly influence the result.
Traceability of measurement results and calibration standards	Using standards to check that the measurement obtained relates to the amount of analyte present and ensuring that this is traceable to some well-defined, recognised standard (see Box 6 Section 2.6)
Precision (repeatibility and reproducibility)	Checking that the method generates results that show satisfactory agreement under the same conditions (i.e. method, operator, laboratory, material) over a narrow time period (repeatability) and under more variable conditions (i.e. same method and material but different operator, laboratory and equipment) over a longer time period (reproducibility) - see Section 2.7
Method bias	Checking that the method is not subject to bias (i.e. consistently over- or under-estimating the amount of analyte present- see Section 2.7)
Method specificity	Checking that the method measures only the analyte of interest (i.e. shows good specificity for that analyte)
Limit of detection and of quantitation	Assessing how small an amount of analyte can be detected and/or quantified
Ruggedness	Assessing how small changes in the method (e.g. temperatures, pH) affect the result
Records / documentation	Documenting all the validation experiments to provide written evidence of how method performance was assessed and of the final method itself
Confirmation of fitness for purpose	Working through a final checklist of activities to confirm that the validation has addressed what it set out to address

Note that these are just examples of the kinds of aspects of a method that need to be explored during validation. Some of the terms are covered in more detail in other sections (and cross-referenced from the table) and also explained in the glossary. Further information and extensive practical advice on approaching method validation are given in the LGC/VAM guide on which this table is based.

Adapted from: LGC/VAM (2003) In-house method validation: a guide for chemical laboratories.

Other systems exist. For example, in the UK the Food Standards Agency publishes an Information Bulletin on Methods of Analysis and Sampling for Foodstuffs, which contains much discussion on approaches to validation as well as studies of validation trials of particular methods. Much of the content relates to methods that are proposed for inclusion in EU regulations and directives or Codex standards (an internationally agreed set of standards covering many aspects of food quality and safety).

Irrespective of the validation system, and however rigorous it might be, it is widely accepted amongst analysts that a laboratory using a method for the first time should ensure that the method works in its hands.

2.5 Quality controls and standardisation

Even if the method being used is validated and known to perform reliably, it is important to adopt measures that will assure acceptable performance on an on-going basis and highlight problems should they arise. One simple way of doing this is to include in the analysis materials known to contain the analyte at a particular level (or levels) or known not to contain the analyte. These materials act respectively as positive and negative controls. These can be 'in-house' materials, designed to reflect the combination of food matrix, analyte and analyte concentration appropriate to the analysis being conducted. In this case, they should be prepared with great care and checked thoroughly - for example for even distribution of the analyte by using analysis of replicate samples. Alternatively they can be commercially available reference materials from a reputable source and with a fully documented history (see Box 5).

Whilst providing a useful check, controls built into the method are only one part of quality control. Others can be regarded as 'prerequisites' in that, if developed and deployed properly as a part of routine laboratory procedures, they will go a long way towards assuring the reliability of the results. Many of these feature in Figure 2 (p12), which provides an overview of the many things that can influence the reliability of an analysis. For example, documenting the full method as a standard, easy-to-follow protocol will reduce the likelihood of mistakes being made, as will

Box 5 - Reference material

Reference materials contain known amounts of the analyte in an appropriate matrix - for example a crab paste containing accurately determined amounts of particular metals. They provide a valuable check that a method is performing as it should as a part of ongoing quality control. They are also extremely useful in method validation. Some reference materials can be created in-house by adding known amounts of analyte of known purity to an appropriate food matrix.

Some are distributed as certified reference materials - that is, they are supplied with a certificate stating the amount of analyte present and an estimate of the uncertainty in the quoted value. A good example of this is provided by ERM® - groups of certified materials that are subject to a rigorous peer evaluation. Because these materials are fully traceable to stated references and pass through a well-defined preparation system, their use instils great confidence in the methods validated through their use.

The materials covered by the scheme span food and agriculture, environmental analysis, health related analyses, and industrial and engineering materials. The scheme also covers a range of compounds certified for purity, concentration and activity (see Box 6 - Reference standards). The materials that fall within the food and agriculture group are classified into the following sub-groups:

- Potable water and beverages
- Animal matter
- Plant and vegetable matter
- Processed food and foodstuffs not covered by the above groups
- Animal feeding stuffs
- GM materials

The ERM® system is a collaborative venture between three major European analytical laboratories: the Institute for Reference Materials and Measurement (an Institute of the European Commission's Joint Research Centre), LGC in the UK, and the Federal Institute for Materials Research and Testing (BAM) in Germany. This further ensures the confidence of analysts making use of the materials.

For further information see: *www.erm-crm.org*

training staff to ensure that they understand what they are doing and why they are doing it - i.e. that they are competent to undertake the task. Carefully calibrating equipment (e.g. balances, pipettes) as part of a routine schedule will help to eliminate bias. Good 'housekeeping' measures such as proper labelling and storage of reagents, for example, will help avoid the introduction of gross errors. Detailed and clear note keeping will help to trace the source of problems (e.g. errors in calculations or dilutions) should they arise.

Box 6 - Reference standards

Reference standards are extremely important for calibrating instruments to obtain reliable quantitative measurements of an analyte. Using a series of standards, containing known amounts of analyte, the analyst is able to relate the signal from the instrument to the amount of analyte in an 'unknown' sample. Many compounds are available as pure materials with a specified purity value, or as solutions of known concentration. These are used routinely by analysts for the calibration of instruments. As with reference materials (see Box 5), standards can be obtained as certified reference materials where accompanying documentation not only specifies the amount present or the purity of the material but also includes an estimate of the uncertainty of the quoted value. Again, the ERM® system provides a good example, as it includes a range of compounds certified for purity, concentration and activity.

For further information see: *www.erm-crm.org*

2.6 Quantitative measurements

Many analyses are quantitative - that is, they enable the determination of the amount of analyte present. This is achieved by analysis of known amounts of the analyte (in standard solutions) so that the response or signal generated by the instrument can be related to the amount of analyte present. With machines that have been calibrated in this way, it is possible to analyse samples for which the amount of analyte is not known and use the signal generated to determine the amount of analyte present. This is depicted in Figure 4.

The principle of this approach is the same for a range of techniques, even though the signal differs between the techniques. In spectrophotometry, for example, the signal is a change in the amount of light absorbed by a compound (see Section 3.6). In chromatography a range of detectors are used to generate a trace of peaks called a chromatogram (see Section 3.3) in which the peak area can be related to amount of analyte present. In each case it is important to use reliable reference standards to calibrate the instrument, as illustrated in Figure 4.

Figure 4 - From reference standards to quantitative measurement

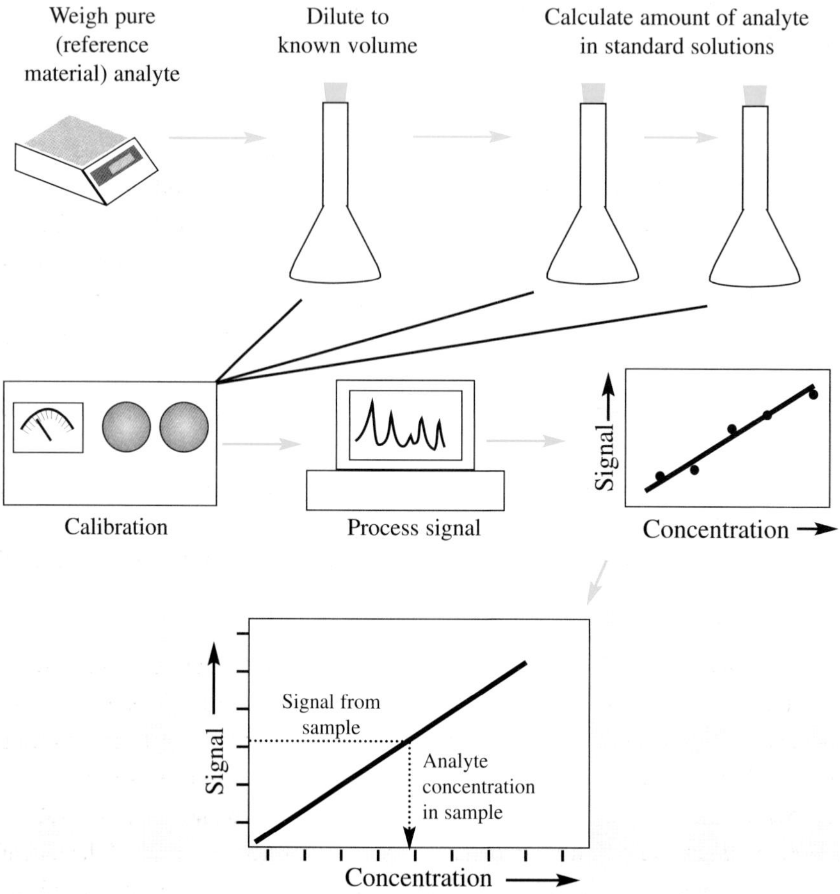

2.7 Measurement uncertainty, its sources and its estimation

No analytical result is perfect - it is a best estimate of the true position. Errors will creep in at various parts of the analytical procedure and these will affect the final result - they will introduce 'uncertainty'. Even repeat analyses of the same sample under tightly controlled conditions and by an experienced analyst will provide slightly different results. The word 'errors' in this context does not mean mistakes - though obviously mistakes will contribute to inaccuracy in the result. The various points at which things can go wrong during an analysis were outlined in Sections 1.3 and 1.4, and the kinds of controls and systems that can be used to prevent mistakes and assure reliable results have been covered subsequently. But even employing all these, the result will be uncertain because some factors that affect the result will be beyond the control of the analyst or are very difficult to control.

This is best illustrated through some examples:
- The material used to prepare a standard solution might not be 100% pure (and any measurement of the level of impurity will itself be an estimate)
- Volume measurements in measuring cylinders, pipettes and volumetric flasks can be slightly inaccurate (e.g. if the meniscus is between two graduation marks, if there is a small difference between the stated volume and the actual volume when the item is filled to the graduation mark, or if the temperature at the time of use is slightly different from that during calibration)
- The display on a weighing balance will only show the mass to a certain number of decimal places
- The method might involve some subjective judgement (e.g. the end-point colour change in a titration)
- The analyte will probably not be dispersed evenly (homogeneously) through the sample

Factors such as these will introduce uncertainty in the final result. By understanding them the analyst will often be able to reduce the uncertainty associated with a result. There are two general sources of uncertainty - random errors and systematic errors. With random errors, the variation from one measurement to another is unpredictable. For example, if the analyte is not distributed in the sample homogeneously, sub-samples will contain differing amounts, leading to different estimates of the overall amount. To compensate for this, a number of sub-samples can be analysed and an average calculated.

In contrast, systematic errors introduce bias so that the results differ from the true result by the same amount each time a measurement is made. For example, if a pipette is consistently dispensing a slightly larger volume of test extract it could lead to a consistent over-estimate of analyte in the test. This type of uncertainty will not be corrected by repeat analyses (as they will all be subject to the same bias). If the source of the systematic error is known and understood, it is sometimes possible to make corrections - in the example given, a 'correction factor' could perhaps be introduced into the calculations.

The measurement uncertainty quoted with a result is a combination of both systematic and random effects. It is usually quoted as a '± value'. For example, from analysis the amount of fat in a sausage might be quoted as 19.5 ± 0.8 g per 100g. This means the analyst believes (based on the uncertainty in the result derived from knowledge of the method of analysis) that the true value is between 18.7 and 20.3 g/100g.

The terminology used in these cases is important as it enables analysts to communicate clearly when problems arise - and to devise ways of dealing with uncertainty. Confusion arises with some of the terms used - particularly 'accuracy' and 'precision' which are related but different (see Box 7). Methods for estimating uncertainty do exist but are beyond the scope of this book. Further discussion can be found in Barwick and Prichard (2003).

The important point is that even with the very best controls in place the analytical result is only a best estimate of the 'true' value - it can never be perfect. But if the analysis is done properly, then allowance can at least be made for uncertainty, so that the result is sufficiently meaningful to use as the basis of a decision. Also, repeat analyses can give a good estimate of the variation in results, which will enable the analyst to determine the number of analyses required to establish whether there is a difference between samples.

Box 7 - Accuracy, precision and bias

To understand the distinction between 'accuracy', 'precision' and 'bias', it is useful to think of an archer aiming arrows at the centre of a target. If all the arrows cluster tightly at the centre of the target, the archer can be described as being accurate. All the arrows are close together so the archer's aim is also precise. In addition, all the arrows are close to the centre of the target (at which the archer was aiming) so there is no bias.

If the arrows are more scattered, but still around the centre, there is no significant bias but the archer is imprecise. If all the arrows hit the target in a cluster but away from the centre, then the archer is precise but showing a bias as the hits are consistently wide of the mark in a particular direction. Similarly, a less precise archer might also show bias with arrows consistently to one side of the target but more widely scattered. From this depiction it can be seen that for the archer's shots to be described as accurate they must fall in a small area (i.e. be precise) which is close to the centre of the target (i.e. be unbiased).

The same is true for an analyst who could be using a technique that shows good precision (generating results that show good agreement from repeat analyses) but which are nevertheless biased (i.e. consistently too high or too low) due to errors in the analytical procedure.

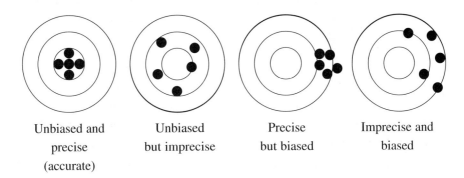

Unbiased and precise (accurate)   Unbiased but imprecise   Precise but biased   Imprecise and biased

2.8 Accreditation

A food company without any laboratory facilities of its own wants some food samples analysed. It could use any of a wide range of laboratories that offer commercial analytical services. But how does it know which one to pick? The first question is perhaps, 'Does the laboratory do the analysis we want?' The second will almost certainly be, 'Is the laboratory accredited?'

Accreditation is recognition of the competence of an individual, laboratory or company by a recognised, independent (third party) authority. Knowing that a laboratory is accredited gives the user assurance that the laboratory operates recognised systems and procedures for quality assurance, and that its use of these is checked (i.e. audited) on a regular (e.g. annual) basis and against a defined standard (the scheme standard). For a non-accredited laboratory, the client does not have this independent reassurance.

The many factors that can influence the reliability of an analysis were outlined in the introduction to this chapter (see Figure 2, p12). These include everything from the maintenance and calibration of equipment, choice and validation of methods, preparation and handling of reagents, and handling of samples, to staff training, laboratory management and quality control procedures. An audit by an accreditation body will consider all these aspects, looking at a sample of them in some detail, and comparing them with the accreditation scheme's standard.

For any aspect of the laboratory's activities that fall below the scheme standard, the laboratory will be issued with a non-compliance (i.e. a requirement to correct the problem). Non-compliances can be minor (i.e. relatively small) or major (more serious). The action that the laboratory takes to address the non-compliance is called a corrective action. The granting of accreditation (or renewal if the laboratory has already achieved accreditation) is conditional upon resolution of any non-compliances. In this way, the certificate of accreditation provides an indication to the laboratory's clients that the laboratory has been independently assessed and deemed to be operating at an appropriate standard (at least at the time of the last audit).

Accreditation is awarded for a defined range of tests on designated materials or sample types; this is called the scope of the accreditation. The laboratory identifies which methods and materials are included in its scope and so subject to scrutiny at the audit. New methods and materials can be added (to expand the scope) and others can be removed. A laboratory accredited for certain methods and/or materials is at liberty to carry out analyses that do not fall within the scope. This is quite common. However, it is not at liberty to claim that it is accredited to perform analyses that fall outside the scope. This would constitute fraud.

There are various accreditation schemes for food chemistry laboratories. Examples include accreditation by UKAS (the United Kingdom Accreditation Service) and CLAS (the Campden Laboratory Accreditation Scheme). Both UKAS and CLAS address the principles and requirements in the internationally recognised standard BS EN ISO/IEC 17025:2000 (*General requirements for the competence of testing and calibration laboratories*).

2.9 Proficiency testing schemes

Even laboratories that have adopted good laboratory practice and achieved third party accreditation should challenge their systems to ensure that they are performing well. One very good way of doing this - because it enables direct comparison with other laboratories and so provides a good indication of relative performance - is through participation in a recognised proficiency testing scheme. Participation in proficiency tests is a requirement of the BS EN ISO/IEC 17025 standard mentioned above.

These schemes typically work by providing participating laboratories with standard samples for the laboratory to analyse 'blind' - that is, the organisers know the level of the analyte but the laboratory does not. The laboratory analyses the sample(s) using its usual method (which might differ from lab to lab) and submits its results to the organiser. The organiser compiles these results (e.g. plotted as a histogram) into a report that also contains the 'true' result and, for each lab, a score (called a z-score) reflecting how close it was to the 'true' value (see Figure 5). The labs are not named but coded and each lab is told only its own code. This means it can compare its result with the 'true' value and with the results obtained by other laboratories. It can also compare its score with other laboratories without confidentiality being breached.

This gives a laboratory good independent feedback of its performance for a particular test.

Various schemes are now well established. Examples include FAPAS® (Food Analysis Performance Assessment Scheme), QM, IMEP® (International Measurement Evaluation Programme) and GeMMA (Genetically Modified Material Analysis Scheme). FAPAS® and GeMMA deal with analysis of food and feed, QM with food and environment analysis and IMEP® with food, environmental, medical and industrial materials (e.g. water, sediment, body fluids, plastics, car exhaust catalysts). To ensure standardisation, the schemes themselves are organised according to internationally agreed rules for proficiency testing schemes.

Figure 5 - Illustrative plot of z-scores from a proficiency test

The figure shows the z-scores obtained by several laboratories in a proficiency test. z-scores between -2 and +2 indicate a satisfactory performance. z-scores between -3 and -2 or between +2 and +3 indicate a questionable result. z-scores outside the range -3 to +3 indicate an unsatisfactory result. Where a laboratory obtains a 'questionable' or 'unsatisfactory' result, it should investigate the cause. Note that the data in this illustration are fictitious and do not relate to real laboratories nor to a real proficiency test.

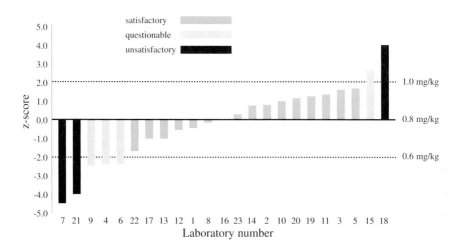

The schemes can cover a wide range of analytes and matrices. In FAPAS®, for example, individual proficiency tests are referred to as 'rounds' and are organised as series. For example, the aflatoxin series might include separate rounds based on maize, chilli, figs and peanuts. Table 6 lists some other examples of FAPAS® series and rounds. GeMMA has covered analysis of various soya and maize flours as well as mixed and processed cereals, for specific genetically modified soya and maize materials. QM is organised as a series of schemes including the dairy chemistry scheme, beverages scheme and meat chemistry scheme.

IMEP® is organised by the Institute for Reference Materials and Measurements (IRMM) which falls within the European Commission's Directorate-General Joint Research Centre. Examples of recent IMEP® food and drink based rounds include metals in tuna, rice and water.

2.10 Choosing a laboratory to do an analysis

This chapter illustrates that there are many aspects to doing an analysis properly. It is not just about the analysis itself, but about maintaining systems, procedures and an infrastructure to make the analysis as reliable as possible. As the examples of uses of food analysis show - in Chapter 4 - some very big decisions rest on the results of food analysis. It is important that the analysis is done properly.

In choosing a laboratory to carry out analyses, a company should ask various questions to reassure itself that the laboratory can provide what is needed. Laboratories vary considerably in what they can provide, not just in terms of reliability of the analysis but also in matters such as interpretation of the results, supporting explanations and advice. With regard to the reliability of the result, questions should include:

- Does the laboratory offer the service / analysis being sought?
- Is the laboratory accredited and by whom?
- What is the scope of that accreditation and does it specifically include the analysis being sought?

Table 6 - Examples of FAPAS® series and rounds

Series	Round (matrix)	Analyte
Fish and meat authenticity	Chicken	Other meat species
	Fish pieces	Species identification
Nutritional	Canned meat	Moisture, ash, total fat, nitrogen, dietary fibre
	Milk powder	Moisture, ash, total fat, nitrogen, lactose
	Dried sauce mix	Moisture, ash, total fat, nitrogen, sodium, chloride
Aflatoxins	Animal feed	Aflatoxin B and G or total aflatoxin
Veterinary drug residues	Pig kidney	Nitrofuran metabolites
	Milk powder	Chloramphenicol
Metal contaminants	Tomato puree	Tin and iron
	Canned fish	Arsenic, mercury, copper and zinc
Soft drinks	Cola	Caffeine, benzoic acid, acesulfame K and saccharin
Fruit juice	Orange juice	Brix, pH, total acidity, total sugars, calcium, potassium and phosphorus
Acrylamide	Breakfast cereal	Acrylamide
Specific migration	Baby food	Semicarbazide
Alcoholic drinks	Rum	Alcoholic strength, methanol and propanol

FAPAS® is the Food Analysis Performance Assessment Scheme. Individual proficiency tests are referred to as 'rounds' with similar rounds grouped as 'series'. For further information see *http://ptg.csl.gov.uk/schemes.cfm* For further information on QM see *www.qualitymanagement.co.uk* and for details of IMEP® see *www.irmm.jrc.be/imep/* Further description of many of these analytes and matrices is given in Chapter 4 where the uses of food analysis are discussed in greater detail.

- Does the laboratory participate in proficiency tests and has it specifically participated in a test for the analysis being sought? If so, how did its performance compare with that of other laboratories?
- How does the laboratory know that the method is fit for the intended purpose? How has the method in question been validated?
- What quality controls do they operate? Do they use reference materials?

The issue of cost should always come after these 'performance' questions. The cost of analyses will vary enormously. Accreditation, participation in proficiency testing schemes, training staff, and ensuring that methods are validated and fit for purpose, all take time and cost money. A laboratory that does not undertake these activities can offer tests at cheaper rates than a laboratory that does undertake these activities. However, if the laboratory does not participate in these activities, its prices might be cheap but users will have much less confidence in the results generated. If the decision is sufficiently important to warrant getting the analysis done, it is sufficiently important to warrant getting it done properly. So, perhaps the final question, and one which the company should ask itself, is 'What are the consequences of getting the wrong result?' If they are insignificant, the analysis is not worth doing. If they are significant, it's worth ensuring that the laboratory takes the right approach to get the right result.

3. TECHNIQUES IN CHEMICAL ANALYSIS

Analysts have at their disposal a wider range of analytical techniques than ever before. And the sophistication of many of these would have been almost unimaginable just a few decades ago. This means that the analyst can now measure lower levels of a wide range of compounds in many different sample types. But it also means that the analyst has to be careful about the approach taken. As discussed in the last chapter, getting the right result requires the correct approach - and this includes using the right method of analysis.

This chapter looks at some of the main techniques used in food analysis and outlines the principles underlying them. It also gives some examples of their use in food analysis to illustrate how the techniques relate to the routine analysis of foods (though this is covered to a much greater extent in Chapters 4 and 5). In some cases it will be apparent that different techniques can be used to detect the same or similar analytes, so it is important to bear in mind that the examples given for each technique are just that - examples. Also, as the intention is to illustrate the uses to which techniques are applied, the level of detail given is necessarily limited. The references cited in Section 8, however, give plenty of leads for finding out more. Taken as whole, this chapter emphasises one of the principles of VAM - namely, the importance of using methods and equipment that are fit for purpose - as well as illustrating that the choice of method is not always obvious.

A method of analysis typically involves several stages, and can involve a combination of techniques. Figure 6 gives a generalised overview of the main stages in an analysis, with examples of where some of the techniques that are covered here fit in.

Figure 6 - The main stages of a method of an analysis

Stage	Examples
Sample receipt	Checking and logging samples (2.2 and Table 4)
Sample pre-treatment	Drying, grinding, blending
Separation of analyte from matrix	Ashing, acid digestion, solvent extraction (see 3.1)
Clean-up of extract	Solvent extraction (Box 8), 'affinity' columns (Box 9)
Measurement of amount of analyte	Analytical technqiues such as chromatography (3.3), spectrophotometry (3.6) and atomic absorption spectrophotometry (3.5)
Calculation and reporting of results	See Section 2.6

The flowchart depicts the main stages in an analytical method. The numbers in parentheses refer to sections where further information can be found on the examples mentioned.

3.1 Sample preparation and analyte extraction

On receiving the sample the analyst has to prepare it for analysis. Initially this will involve checking that the sample is appropriate for the specified analysis, is available in the right amount and the right condition, and so on. In some cases, only limited further preparation is required - see, for example, moisture analysis below. In other cases the analyte might have to be extracted and the extract then cleaned to remove compounds that would interfere with the analysis. This is often particularly the case for trace analytes (i.e. low levels of the compound of interest) in complex matrices, which can require extensive preparation and clean up (e.g. pesticides, vitamins).

One approach that is commonly used is solvent partitioning, which involves separating components of an extract from a sample on the basis of whether they are more soluble in water or organic solvents like diethyl ether or petroleum ether. A good example is in the clean-up of extracts for vitamin E analysis (see Box 8). Other analyses can be simplified by use of specific clean-up procedures - for example, aflatoxin analysis uses single-step affinity columns which 'pluck' the aflatoxin from a mixture whilst discarding the unwanted components of the extract (see Box 9). These are just two of the many variations in sample clean-up, but they illustrate the point that the exact procedure adopted will require careful consideration of the chemical nature of the analyte and the chemistry of the food matrix.

A further point that the analyst will need to bear in mind is how the extract will be analysed. In the case of fat analysis, for example, some of the treatments used to release fats attached to protein within the food (and necessary for total fat analysis) limit further analysis of the fatty acids in the extract. So, sample preparation should be seen as an integral and important part of the analysis.

Box 8 - Solvent partitioning

Many analyses begin with the extraction of the analyte from the food sample into a solvent. Analytes that are soluble in water are usually extracted into water-based solvents or solvents that mix with water (called 'polar' solvents). In contrast, fat-soluble analytes are usually extracted in organic solvents like petroleum ether or chloroform (called 'non-polar' solvents). Many of the compounds extracted in this way might be soluble in both types of solvent to a greater or lesser extent, and this can be used to 'clean-up' the extract (i.e. isolate the analyte from other unwanted components).

continued....

One example is in the extraction of vitamin E from cereals, for analysis by HPLC (see Section 3.3.2). The cereal is initially extracted with a mixture of ethanol and alkali which is then mixed with petroleum ether with the addition of more water. The entire mixture is allowed to settle out into two layers: an aqueous layer, which contains the water-soluble material, and the petroleum ether layer, which contains the vitamin E and other fat-soluble material (see figure below). This is called solvent partitioning because the components of the extract partition themselves between the liquid phases.

The differential solubility of analytes in solvents is an important feature not just of extraction and clean-up but also separation of compounds by chromatography (see Section 3.3).

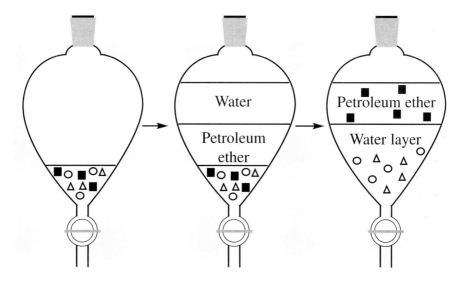

The cereal is blended in ethanol/alkali to extract various components (unfilled shapes) as well as the vitamin E (shaded squares). This extract is then mixed with petroleum ether before further water is added. The water and original ethanol/alkali layers mix to form a large aqueous phase (water layer), on top of which floats the petroleum ether, which is immiscible with water. The alkali helps to make some of the small fatty molecules (but not vitamin E) more soluble in water. The vitamin E, being more soluble in the petroleum ether than water, moves into the upper layer, leaving many of the other components in the aqueous layer.

Box 9 - Affinity column clean-up

Extracts can be cleaned to remove unwanted compounds but retain the analyte, through the use of affinity columns. This approach is used routinely in aflatoxin analysis for example. The sample is extracted and passed through a column that contains antibodies to the aflatoxin. These antibodies recognise and trap the aflatoxin while the other components of the extract pass by. The aflatoxin can then be washed from the column with a solvent for further analysis (e.g. by HPLC). The use of antibodies as an analytical tool is described further in Section 3.12 and in much more detail in Jones (2000).

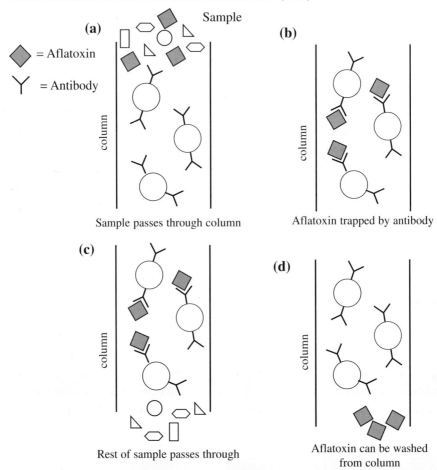

The sample is passed through the column (a). Antibody immobilised on the beads in the column traps aflatoxin (b) as the sample passes through the column. When all the other sample components have washed through the column (c), the aflatoxin can be washed off the antibody (d), and analysed separately.

3.2 Gravimetric methods

Gravimetric methods are relatively straightforward in principle - the material of interest is extracted and weighed. The weight obtained can then be expressed in whatever way is appropriate - for example, in g/100g product, as percentage weight by weight (% w/w) and so on. Examples of applications of gravimetric determination in food analysis are listed in Table 7 and include ash, fat, global migration, moisture and specific gravity of liquids.

Although the approach is simple in principle, in practice it can involve complications. In the case of fat content, for example, the method by which the fat is extracted from the food affects the amount obtained (so-called 'free' and 'total' fat). It is important that the analyst appreciates which needs to be determined and why (see Section 5.5 for further discussion of this). In the case of moisture analysis, the result obtained by the gravimetric approach can differ from that obtained by the Karl Fischer method (based on titration - see Section 3.8). For example, moisture determination in products such as sugar confectionery and molasses, which are prone to decomposition at high temperatures, tend to require lower temperatures for drying or are done by the Karl Fischer method (Kirk and Sawyer, 1991). Similarly, in some products volatile components such as alcohol (e.g. in paté with whisky), volatile oils (e.g. in herbs and spices), or acetic acid (e.g. in fruit mince meat tart) would be lost with water on heating of the sample. In these examples, it is important to be clear about why the material is being analysed and what needs to be determined, when deciding which approach to take. That is, carrying out the analysis to meet an agreed need (VAM Principle 1).

Another factor that could easily be overlooked but which could have a significant effect on the result is ensuring that the balance used for weight determination is calibrated properly. This should involve the use of traceable standard weights, an example of which is given in Box 10. This can help ensure that measurements made in one location are consistent with those made elsewhere (VAM Principle 5).

Box 10 - What is a kilogram?

If an analyst in a test laboratory wants to weigh 1 kg of a material, they would probably use an appropriate weighing balance. But how would they know that the balance was accurate - that the weight displayed was correct? They would have much more confidence in the balance reading if the machine was well maintained, regularly serviced, and the calibration frequently checked and adjusted when necessary.

To calibrate the balance a series of calibrated standard weights (including a 1 kg weight in this case) are placed on the balance and the balance reading is recorded. If necessary, the balance is adjusted to make sure that the balance display agrees with the mass of the calibrated weight. To ensure that the 1 kg standard weight used for calibration is accurate, it would be checked against a standard weight held by an appropriate calibration laboratory (for example the National Weights and Measures Laboratory in the UK). Their 1 kg standard, in turn, would have been checked against a national standard, which in the UK is at the National Physical Laboratory. And this standard would have been referenced against the

continued....

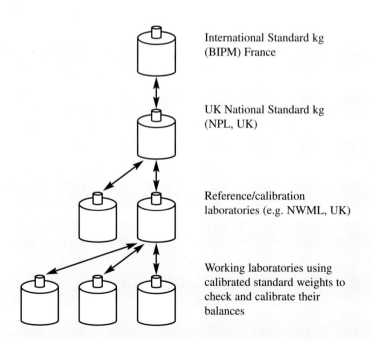

International Standard kg (BIPM) France

UK National Standard kg (NPL, UK)

Reference/calibration laboratories (e.g. NWML, UK)

Working laboratories using calibrated standard weights to check and calibrate their balances

internationally agreed definitive standard, the international prototype kilogram, held at the International Bureau of Weights and Measures (BIPM) at Sèvres, France. This is called a 'chain of traceability' (see the figure opposite).

In each case the standard weight is traceable back to the securely held definitive standard. This concept of traceability of standards is extremely important in analysis, not just for weights but for a range of reference materials (see Sections 2.5 and 2.6). It is also important in fulfilling VAM Principle 5 'Analytical measurements made in one location should be consistent with those elsewhere', in that it enables laboratories to ensure they are comparing like with like. By using weighing balances which have been properly calibrated with traceable standard weights, an analyst can be confident that 1 kg of material weighed on one balance will be the same amount of material as 1 kg weighed on a different balance.

Table 7 - Examples of uses of gravimetric methods in food analysis

Analyte / Parameter	Comments
Ash	Organic matter is burnt off and remaining material (ash) gives an indication of the total mineral salt content of the food
Global migration (from packaging into olive oil)	Packaging is weighed, soaked in olive oil (for migration compounds to leach out), dried and re-weighed. See Section 5.2
Moisture	Water is driven-off a weighed sample by heat and the residue re-weighed to determine water content by difference
Specific gravity of liquids (e.g. for alcohol determination)	Compare weight with that of equivalent volume of water (and consult published tables to relate specific gravity to alcohol content)
Fat	Fat is extracted in solvent which is then evaporated off to leave the fat, which is weighed

3.3 Chromatography

Chromatography is the separation of components of a mixture on the basis of differences in their preference (e.g. solubility, adsorption) for either of two phases. This is similar to the principle that applies in solvent partitioning, as described in Box 8 for the clean-up of an extract containing vitamin E. However, in chromatography one of the phases (the mobile phase) moves, washing over or through the other (stationary phase), and this enables complete separation of components from complex mixtures.

Separation alone, whether by chromatography or other means, is not enough: once the components have been separated they have to be detected and/or quantified with an appropriate system. This section looks first at the general principle of chromatographic separation, then at some examples of different separation systems (e.g. gas chromatography, liquid chromatography) and then considers some of the detection systems available.

Chromatography is a science in its own right and there are many more variations on the technique than described here. Wilson and Walker (2000) provide a good explanation of these. The power and versatility of chromatography, founded on these variations, means that it has found many uses in food analysis. Some examples are given in Table 8.

In a chromatographic system, the stationary phase can be a solid, a liquid or a gel (or a mixture of these) while the mobile phase is a liquid or a gas. The mobile phase flows across or through the stationary phase, sweeping components of the mixture with it. Those components with more affinity for the mobile phase tend to stay in that phase and get moved along relatively quickly. Those components with more affinity for the stationary phase tend to stay in that phase and so move along relatively slowly (see Figure 7). The basis of separation between the two phases can include solubility, charge, size or affinity for a specific trap such as an antibody (see Box 9, p46). The stationary phase can be thinly coated on a flat plate (in thin layer chromatography - TLC) or, more often, is packed into a column (in column chromatography such as GLC and HPLC - see Sections 3.3.2 and 3.3.3).

Table 8 - Examples of uses of chromatography in food analysis

Analyte	Typical samples	Method
Aflatoxins and ochratoxin	Cereals, nuts, dried fruit, coffee, spices	HPLC - with fluorescence detection
Aspartame, acesulfame-K, saccharin (sweeteners)	Soft drinks	HPLC - with UV detection
Benzoate and sorbate (preservatives)	Soft drinks, desserts	HPLC - with UV detection
Milk fat	Milk, milk powder, cheese	GLC - with flame ionisation detection of butyric acid derivatives
Fatty acids	Oils and fats	GLC - with flame ionisation detection
Flavour and taint compounds	Raw materials, ingredients, processed foods and beverages	GLC - with various detectors such as electron capture and flame photometry
Nitrate and nitrite	Fruits, vegetables, cured meats	HPLC - with UV detection
Pesticides	Fruits, vegetables, cereals	GLC with detection by mass spectrometry, nitrogen-phosphorus detection or flame photometry
Sugars	Raw materials, ingredients and processed foods	HPLC with refractive index detection
Vitamin A	Milk powder	HPLC - with UV detection
Vitamin C (ascorbic acid)	Fruit juices and soft drinks	HPLC - with UV detection
Vitamin D3	Milk and milk products	HPLC - with electrochemical detection
Vitamin E	Cereals	HPLC - with electrochemical detection
Water soluble colourings	Confectionery, soft drinks	TLC (identification), HPLC with UV detection (quantification)

GLC - gas liquid chromatography; HPLC - high performance (pressure) liquid chromatography; TLC - thin layer chromatography. Note that the detection systems mentioned are described in Section 3.3.4 (p58).

Figure 7 - Principle of separation by chromatography

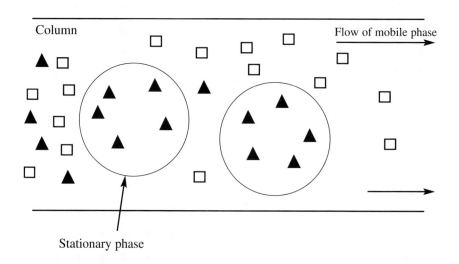

Some components (□) 'prefer' the mobile phase and so move along quickly. Others (▲) prefer the stationary phase, and so move along more slowly. This results in the separation of the components of a mixture.

The following sections look at three different techniques - TLC, GLC and HPLC - which are routinely used in food analysis. The choice of which approach to use is not always straightforward. TLC can be useful for identification of compounds but not for quantification. Both GLC and HPLC can be used for identification and quantification. GLC tends to be used for volatile compounds (or compounds that can be made volatile easily). HPLC is very versatile, because of the many combinations of phases available, and so can now be used for many food analytes.

3.3.1 TLC (thin layer chromatography)

In TLC, a solid phase (e.g. silica) is coated on a support (e.g. a glass plate). The sample is applied to the plate near one end, and this edge of the plate is stood in a reservoir of mobile phase. This can be any or a mixture of a wide range of solvents. The mobile phase moves up the solid phase by capillary action. As it passes over the sample, it carries the components with it and they become separated on the basis of their preference (relative affinity) for the stationary and mobile phases. The extent

Figure 8 - Stylised outcome of TLC separation of two food colours

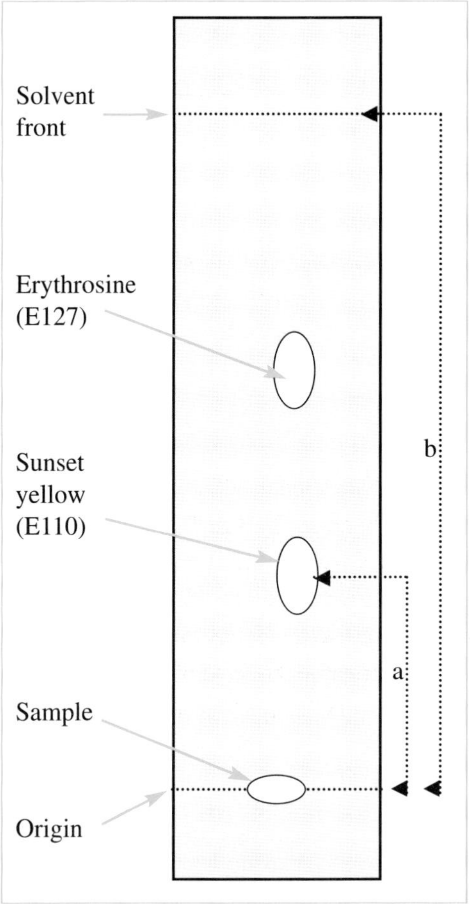

The sample is applied at the origin and the TLC plate is stood, origin-end lowermost, in a pool of solvent (mobile phase). The mobile phase moves up the plate by capillary action. As it moves over the sample it carries components of the sample with it. Those components with a preference for the mobile phase (e.g. erythrosine) will tend to stay in this phase and move quickly. Those with a preference for the stationary phase (e.g. sunset yellow) will tend to stay with this phase and so be moved along more slowly. The distance travelled by each component can be expressed in terms of the Rf - the distance it travels from the origin relative to the distance travelled by the solvent. In this example the Rf for sunset yellow is a/b.

to which a component moves from the site of application (origin) is expressed in terms of its Rf (retardation factor). This is the ratio of the distance moved from the origin by the analyte to the distance moved from the origin by the solvent (the solvent front).

This is a simple and older form of chromatography, which is now used in food analysis to a lesser extent than it used to be. However, one good example of where it is still used is in the separation and identification of water-soluble colours from foods (Figure 8).

3.3.2 HPLC

HPLC stands for high performance (or high pressure) liquid chromatography. It is a form of column chromatography in which the solid phase is a mass of small particles packed tightly into a stainless steel tube. This gives the solid phase a very high surface area so that the components of the sample have much more exposure to the interface of the mobile and stationary phases than in TLC. This allows much better separation (resolution) of the components in a sample (hence the term 'high performance'). However, it also means that much higher pressures are required to force the mobile phase along the column (hence the 'high pressure' in the original name of the technique).

Further variation can be introduced through the mobile and solid phases. For example, some solid phases tend to retain water-soluble compounds while others retain fat-soluble compounds (and are called reverse phase systems). Many different solvents can be used as the mobile phase. Also, the composition of the mobile phase can be altered during a chromatographic run. An example of this is with HPLC analysis for the mycotoxin DON (deoxynivalenol), where the mobile phase is initially a methanol/water mixture but the proportion of methanol is increased until the column is being washed with 100% methanol.

As the components of the sample are washed off the end of the column they pass in front of a detector (e.g. UV spectrophotometer - see Section 3.3.4) and the signal is plotted as a series of peaks on a chart or 'trace' (see Figure 9). The resulting trace is called a chromatogram. In the case of column chromatography, the concept of Rf (see TLC above) is replaced by one of retention time - i.e. how long it takes the component to pass along the column. The peak areas are compared with those of reference standards to determine the amount of analyte.

The performance and versatility of HPLC - based on the columns, solvents and gradients used - means that it has found many applications in routine food analysis. Some of these are listed in Table 8 and a typical HPLC trace (chromatogram) is depicted in Figure 10.

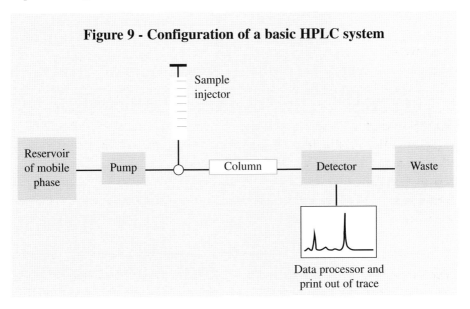

Figure 9 - Configuration of a basic HPLC system

3.3.3 GLC/GC

GLC stands for gas-liquid chromatography. In traditional GLC, the mobile phase is a gas (e.g. nitrogen, helium, argon) which passes over a liquid phase (traditionally a high boiling point liquid such as silicone grease) coating an inert granular support. The term GC (gas chromatography) is broader in that it also includes systems in which the stationary phase is solid. In many modern systems the 'column' is a coil of a very fine capillary tube (up to 50 metres long) with the stationary phase coated on the inside of this tube. The components of the sample move between the two

Figure 10 - HPLC trace of aflatoxins

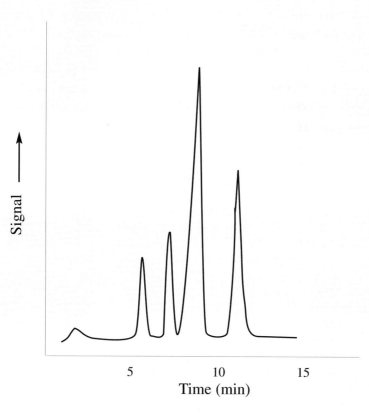

Typically, the aflatoxins are extracted from the food and the extract is cleaned-up (e.g. filtering, then an immunoaffinity column - see Box 9, p46), before being injected on to the HPLC column. They are detected by a fluorescence detector (see Section 3.3.4) as they emerge from the column. The approximate retention times for the four afltaoxins (B1, B2, G1, G2) in this trace are: G2 - 6 minutes; G1 - 8 minutes; B2 - 9 minutes; B1 - 11 minutes. The aflatoxins get their name from the colour they fluoresce (B for blue and G for green). The peak areas can be calculated and compared with those of known amounts of standards to determine the amount of each aflatoxin.

phases on the basis of their relative solubility in each and are eventually carried off the end of the column where they pass a detector, creating a measurable signal (see Section 3.3.4). This is used to create a trace (chromatogram) similar to that obtained in HPLC. The basic configuration of a GC system is shown in Figure 11.

The extract has to be kept in the vapour phase (to allow the components to move into the gas carrier) and so the column is operated at high temperatures (e.g. around 200-250°C for fatty acid analysis). This also speeds up the interactions between the extract components and the phases. Further variations can be achieved with the phases used, temperature gradients during the chromatographic run, column length and the way in which the column is packed, and so on. These variations give GLC much versatility, which means that it has found many uses in food analysis, some of which are given in Table 8 (p51). GLC has proved particularly useful for the separation of fat-soluble compounds, though is by no means restricted to this. Some analytes are not analysed directly but are first converted to more volatile forms.

Figure 11 - Configuration of a basic GLC system

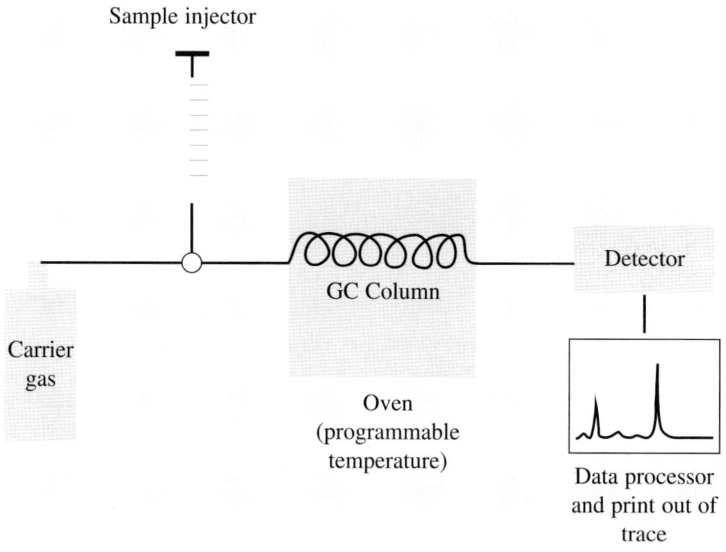

3.3.4 Detection systems

Chromatography enables the separation of components of a mixture, but these then have to be detected in some way. A variety of detectors are available for use with HPLC or GLC (see Table 9). The choice largely depends on the chromatographic system (e.g. presenting the analyte in a liquid or gaseous phase) and the properties of the analyte (e.g. absorbs light, electrical properties). The main detection systems are described below:

- Many compounds absorb ultraviolet or visible light and this is widely used as the basis of an HPLC detector (it is also the basis of UV/VIS spectrophotometry - see Section 3.6). Light passes from a source through substances separated from the sample, to a detector, which is able to measure the amount of light absorbed. In such systems the wavelength can be varied (to select that which is maximally absorbed by the compound of interest). In diode array systems, the absorbance at many wavelengths is measured in a very short space of time (e.g. one hundredth of a second), using a bank of light sensors, to produce a 3-D trace (wavelength, time and peak intensity).

- Fluorescence detectors are also used with HPLC - after being separated by chromatography, the analytes of interest are excited by, for example, UV light. This causes them to fluoresce, and measurement of the amount of fluorescence allows the amount of analyte to be determined.

- Electrochemical detectors are increasingly being used with HPLC, to detect analytes which undergo oxidation or reduction reactions, as they can cause electrical changes when they pass between two electrodes. Different conditions (channels) can be selected that detect particular groups of chemicals and not others - the HPLC method for Sudan dye (see Box 4 p23) involves such a system.

- Flame photometry - used with GLC, this is based on the measurement of visible light of characteristic wavelength, emitted when an element is burned in a flame. It is also sometimes known as flame emission spectrophotometry (see Section 3.5).

Table 9 - Examples of detectors used with chromatographic systems

Detector system	HPLC	GC
Diode array	✓	
Electrochemical detection	✓	
Electron capture detection		✓
Flame photometric detection		✓
Flame ionisation detection		✓
Fluorescence	✓	
Mass spectrometer	✓	✓
Nitrogen-phosphorus (thermionic) detection		✓
Refractive index detection	✓	
UV / visible wavelength absorption	✓	

✓ indicates that this combination of chromatography and detector is routinely used. These detectors are used in a range of chromatographic analyses of foods (see Table 8 p51)

- Flame ionisation - this is widely used with GLC as it responds to a wide range of organic compounds. Components coming off the column are ionised in a flame, changing the electrical properties of the gas between the flame and electrode, resulting in a signal on the detector.

- The nitrogen-phosphorus detector, also used with GLC, is similar to flame ionisation except that the electrode contains a salt crystal (e.g. sodium, rubidium), an arrangement which gives a much higher signal for compounds containing nitrogen or phosphorus than for compounds containing neither of these elements. This is also called thermionic detection.

- Another system used with GLC is electron capture, which is applicable to compounds that capture electrons - for example, compounds containing chlorine such as organochlorine pesticides. As the carrier gas leaves the column it is ionised by a radioactive beam (emitted from a sealed radioactive source), and the electrons thus released carry a current between two electrodes. As the electron capturing compound comes off the column it captures some of the electrons, causing a drop in current which is read as a signal.

Most of these detection systems provide information on the retention times of the components of the extract. Also, by comparing the areas of the peaks obtained for extract components with the areas obtained for known amounts of standards, it is possible to calculate the amounts of the different components (see Figure 4 p32). However, there is always the possibility that two compounds with the same retention time leave the column at the same time (i.e. they co-elute). Mass spectrometry provides much more information than the other systems. It allows the analyst to check that the peak is what they think it is and, because of its importance in chemical analysis, is covered separately and in more detail below.

3.4 Mass spectrometry

In crude terms, mass spectrometry (MS) can be likened to a controlled explosion in which a molecule is blown apart and identified from the characteristic fragments it forms. More accurately, it involves ionizing the molecule of interest and separating the ions on the basis of their mass/charge ratio. An ion formed from the unfragmented (parent) molecule is called the parent ion (or molecular ion), and those from fragments of the molecule called daughter (or product) ions. Depending on the type of mass spectrometry, a particular molecule will generate a fragmentation pattern from which its structure can be elucidated and its identity confirmed.

Traditionally MS was used as a detection and identification system for compounds separated by GLC (and called GC-MS) but innovations in sample presentation and interfacing have led to its use with liquid chromatography (LC-MS) and with direct introduction of sample to the mass spectrometer. This is leading to a proliferation in the applications of MS to food analysis. However, mass spectrometers are still relatively expensive, and so their use is still limited to larger or more specialist laboratories. The following provides a generalised overview.

The process of mass spectrometry can be considered in 4 stages: interface, ionisation of the molecules, separation of the ions, detection of the ions (Figure 12). Typically the sample is introduced into the mass spectrometer through an interface. This allows the sample to be fed from a GC via a separator (that removes much of the GC carrier gas) or from an LC system via a spray to create tiny drops of solvent that quickly evaporate, leaving the analyte.

The analyte is then ionised - typically by bombardment with a beam of electrons, though other systems such as chemical ionisation, light (photo) ionisation or electric field ionisation are also used. The ions are accelerated through the mass analyser, which separates them - for example, in a magnetic field, which deflects the ions of particular mass/charge towards the detector. Traditionally the detector is a surface with which the ions collide and either donate or accept electrons (depending on their charge) causing a (detectable) electrical change in the surface, though a range of systems now exist (see Downard, 2004).

There are many variations on mass spectrometry. Some systems allow selective filtering or trapping of ions, though with some compromise in sensitivity. Being comparatively cheap, these lend themselves to bench top systems, widening the use

Figure 12 - Typical sequence of events in mass spectrometry

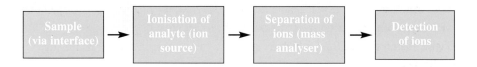

The sample is introduced into the mass spectrometer, for example from a GC or LC separation, via an interface. Molecules present are ionised and the ions separated on the basis of their mass/charge ratios (e.g. in a magnetic field) before detection. Analysis of the parent ion (an ionised form of the intact molecule) and of the daughter ions (ions formed from fragments of the molecules) allows determination of the molecular structure of the analyte and confirmation of its identity.

Table 10 - Examples of applications of mass spectrometry in food analysis

Analyte / Parameter	Purpose
Algal toxins	Safety assurance (these compounds are accumulated by shellfish and so can cause food poisoning)
Allergen proteins (e.g. peanut)	Allergen detection (safety) and understanding allergen structure
Bacterial peptides and toxins	Identification of bacterial spores (e.g. *Bacillus*) and toxins (e.g. *Staphylococcus*)
Disinfectants / cleaning compounds	Developing methods to screen for residues
Environmental contaminants (e.g. dioxins and PCBs)	Monitoring levels in relation to safety and dietary intake
Enzyme (and other protein) profiling in raw materials	Understanding their role in quality and differences between raw materials
Flavour compounds in raw materials, foods and soft and alcoholic drinks	Understanding their contribution to quality and product characteristics
Mycotoxins	Compliance with legislative limits and product safety assurance
Organic contaminants and adulterants (e.g. acrylamide, sudan dyes)	Safety and quality assurance
Pesticide residues (multi-residue screening)	Compliance with maximum residue levels
Veterinary residues (multi-residue screening)	Compliance with maximum residue levels and safety assurance

The examples given include some that are increasingly used as part of 'routine' food analysis (e.g. veterinary residue, pesticide and mycotoxin analysis) where the approach often allows simultaneous determination of many compounds. Some applications are more research oriented, and are aimed at understanding the chemistry of the food and how this affects product quality in its broadest sense (e.g. flavour formation). With new developments in mass spectrometry, and increasing accessibility of the technique, its use in food analysis is likely to increase significantly in the next 5-10 years.

of mass spectrometry. Some of the different systems now in use are outlined in Box 11. Further explanations of the main principles and terminology of mass-spectrometry can be found in Downward (2004) and Wilson and Walker (2000).

Overall mass spectrometry allows the analysis of a wide range of molecules, as reflected in the kinds of applications to which MS has been put (Table 10). This includes the detection of toxins, allergens, and a range of contaminants including pesticide, veterinary, cleaning and environmental residues. However, applications to food enzyme analysis and flavour compound analysis illustrate application of the technique to quality issues as well as the safety issues listed. In chemical terms, perhaps a particularly dramatic illustration of the immense power of mass spectrometry, is its ability to accurately determine the molecular size even of large molecules such as proteins, for example to 100,000 ± 1 molecular weight units.

Box 11 - Mass spectrometry: some variations

Some of the commonly encountered terms used to describe different approaches in mass spectrometry include:

- TOF or time of flight MS - this makes use of the fact that different ions will travel at different speeds and so can be distinguished in analysers which measure the time taken for the ions to travel a given distance.

- Quadrupole systems - these are relatively lightweight mass analysers, which form the basis of many bench-top spectrometers. The ions pass between four cylindrical rods (diagonally opposing positives and diagonally opposing negatives) which causes the ions to follow complicated trajectories with only some making it through to the detector. Because some ions are lost, the system is often referred to as a quadrupole mass filter.

- Tandem MS or MS-MS - a system in which two mass spectrometers are coupled in sequence so that a particular mass spectrum peak (ion) can be subjected to further degradation (e.g. by collision with a gas) and fed into a second analyser which gives further information. If this is repeated for a succession of first peaks, a vast amount of information can be obtained. In the 'triple quadrupole' system, the first quadrupole acts as a mass spectrometer from which a particular peak is fed into the second, which encourages collisions to generate further ions that are then analysed in the third.

3.5 Atomic emission and absorption spectrophotometry

If a sample is vaporised in a flame, atoms of certain elements present in the sample tend to absorb or emit light. The amount emitted or absorbed is related, under standard conditions, to the number of atoms present and, if measured, can be used to quantify the element in question. This approach is used widely in detecting and quantifying the levels of various metals in foods - including sodium, potassium, aluminium, iron, copper, zinc, manganese, tin, chromium, nickel, calcium, magnesium, lead, and cadmium - as well as metalloids such as arsenic, antimony and selenium. (Metalloids are compounds which fall between the metals and non-metals in the periodic table of elements and consequently have some metal-like properties but do not behave entirely like the metals - for more on this see Reilly, 2002). Foods are analysed for metals and metalloids either because they are important nutrients or because they are toxic (or in some cases, as with selenium, both - depending on the amount present in the foodstuff).

Some metals, such as sodium and potassium, emit light at visible wavelengths when vaporised in a flame and so can be measured by atomic (flame) emission spectrophotometry (FES). Many metallic atoms, however, are better detected by measuring the extent to which they absorb light when vaporised in a flame and so are measured by atomic absorption spectrophotometry (AAS), which is also often used for sodium and potassium.

To carry out AAS, the sample is generally ashed in a furnace, and the ash dissolved in acid and then diluted with water. The resulting solution is sucked through a narrow tube into the AAS where it is sprayed into the flame. The temperature of the flame is controlled by the combination of fuel and oxidant gas (e.g. burning acetylene in air creates a flame around 2,500°C). Light of the appropriate wavelength is passed through it and the amount of light absorbed by the sample recorded and compared with that of standard solutions. The details of sample preparation vary with the type of food and metal, and each metal has its characteristic excitation wavelength (i.e. the wavelength of the light that it absorbs most). Figure 13 illustrates the process.

Figure 13 - Schematic of a typical AAS system

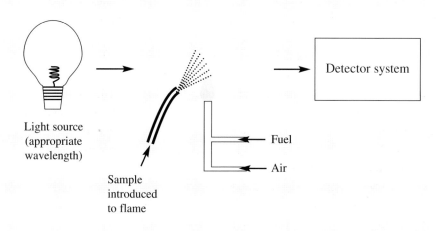

Light of an appropriate wavelength passes through the flame. As atoms from the sample are heated in the flame they absorb some of the light, and this change in light levels is detected.

The atomic emission technique has been taken further with the development of inductively-coupled plasma atomic emission spectroscopy (or ICP-AES). This sophisticated instrument replaces the combustion flame with a plasma flame by using an electrical power source to heat a gas (hence the term 'inductively coupled plasma'). It also uses advanced electronics, optics and multiple detectors. Although expensive, it enables the simultaneous determination of 20 or more elements and so has advantages of high throughput and economy of scale.

ICP has also been coupled to mass spectrometry (ICP-MS) to provide an extremely powerful approach to metal analysis - with detection limits at ng/l, around 1000 times lower than those of AAS. However, this instrumentation is extremely expensive and so used only in a few research or very specialist analytical laboratories, though through economy of scale this has proved useful in surveillance exercises (see Table 22, p105).

3.6 Colorimetry/spectrophotometry

Colorimetry and spectrophotometry both use the absorption of light by an analyte to measure its concentration in a solution. In colorimetry, visible light is passed through a sample and detected by a light-sensitive cell. The amount of light absorbed by the sample can be determined by comparison with reference standards. Within certain limits, this can be related to the amount of analyte in the sample: the more coloured material (analyte) in the sample, the more light is absorbed (see Figure 14).

Figure 14 - Relationship between absorbance and concentration of analyte in colorimetry

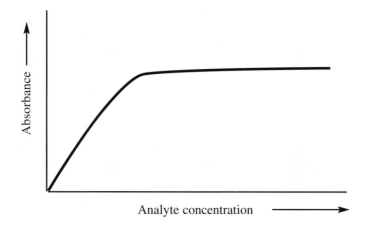

At lower levels of analyte the amount of light absorbed increases linearly with the amount of analyte present, though at higher analyte levels this relationship breaks down. The concentration range over which the relationship is linear is used in analysis. The relationship is encapsulated in the Beer-Lambert Law which states that absorbance is proportional to the concentration of the absorbing substance and the thickness of the layer (i.e. length of the light path through the sample - which is therefore standardised in spectrophotometers).

Spectrophotometry is based on the same principle but the wavelengths involved extend to either side of the visible region - from the ultraviolet (around 190 nm) through the visible region and up to near-infrared at about 1000 nm. Spectrophotometry is also more sophisticated. For example, by using deuterium lamps (for UV) and tungsten lamps (for visible), with sophisticated optical systems, light of narrow bands of wavelength (e.g. 2-10 nm) can be achieved. This can allow better discrimination between compounds with close but distinct absorption maxima. Modern spectrophotometers also allow more sophisticated processing of the signal, scanning a sample over a range of wavelengths, monitoring of a reaction over time, and so on.

Table 11 - Some examples of the use of spectrophotometry in food analysis

Analyte	Comment
Sugars (e.g. lactose, galactose, maltose, sucrose, glucose, fructose)	Enzyme test kits with spectrophotometric determination of product formed (see Box 12)
Starch	Starch is hydrolysed to glucose which is detected as above
Hydroxyproline	Released from collagen by hydrolysis and reacted with a reagent to form a red compound which is quantified spectrophotometrically
Phosphorus	Extracted from ash and reacted with a reagent to form orange-yellow colour which is determined spectrophotometrically
Olive oil quality	UV absorbance by fatty acids
Dietary fibre (Englyst method)	Non-starch polysaccharides are digested to their component sugars which are determined spectrophotometrically (as above)

As described in the text, some analytes can be measured directly on the basis that they absorb light (e.g. olive oil) whereas others are determined indirectly via a chemical reaction that generates a light-absorbing end product (e.g. hydroxyproline, sugars).

Box 12 - Determination of lactose in milk

The level of lactose in milk is determined indirectly. The procedure involves two enzyme-driven reactions. These result in the formation of a compound (NADH) which absorbs UV light and so can be measured spectrophotometrically. The amount of NADH can be related to the amount of lactose.

In the first reaction, the lactose is digested to its constituent sugars, glucose and galactose:

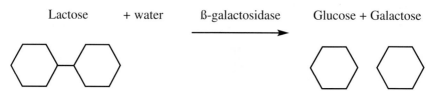

Lactose + water —ß-galactosidase→ Glucose + Galactose

The enzyme β-galactose dehydrogenase then catalyses a reaction in which galactose is converted to galacturonic acid, and NAD^+ (added to the reaction mix) is converted to NADH:

Galactose + NAD^+ —ß-galactose dehydrogenase→ Galacturonic acid + NADH + H^+

The increase in NADH is measured by the absorption of UV light at 340 nm and related back to the amount of galactose in the sample.

This approach will detect two things: any free galactose in the milk plus the galactose released from lactose. As a check, therefore, a portion of the sample that has undergone only the second of the two reactions can be used to detect any free galactose in the milk sample. If necessary this can be deducted before the calculation of lactose. This is a good illustration of VAM Principle 1 - being clear about what the test measures and how the result should be interpreted.

In some cases the light absorption is a property of the analyte itself. For example, olive oil absorbs light in the UV region. This can be used to give an indication of the amount of polyunsaturated fatty acid present, which in turn indicates the oil's quality. In other cases the light is absorbed by the end-product of some reaction, and this is related back to the analyte of interest by way of calculation. Hydroxyproline, for example, is an amino acid found in connective tissue and is used as a marker for the amount of such tissue in meat products (e.g. associated with rind, skin and gristle). The meat proteins are hydrolysed to release the hydroxyproline, which reacts with appropriate reagents to form a red solution. The intensity of the colour (absorbance) is determined by spectrophotometry using light of a wavelength of 558 nm, and this is used to calculate the amount of hydroxyproline in the sample. Another example where the analyte is measured indirectly (via a reaction product) is with sugars. The example of determination of lactose and galactose in milk is outlined in Box 12.

3.7 X-ray microanalysis

When bombarded with a beam of electrons (as happens in a scanning electron microscope), some materials, including glass, give off X-rays. The energy of the X-rays depends on the chemical elements in the sample. Picked up by a detector, the signal can be processed into a count of X-ray hits which can be displayed by a computer as a graph of signal (counts) against energy level (element) as illustrated in Figure 15. Although glass is mostly made of silicon and oxygen, the presence and amount of other elements varies with the type of glass. This provides an elemental profile or 'fingerprint' of the sample, which can then be used to identify the type of glass from which the sample originated. The elemental profile resulting from X-ray microanalysis of a glass sample can be compared with those of other glass types held in a database, to identify its type. For example, container glass has a different elemental composition from Pyrex.

This technique is used to identify glass fragments found as foreign bodies in food (i.e. pieces of undesirable solid material such as bone, fruit stalk or metal). Although all foreign bodies are of concern to food companies, who take stringent measures to try to prevent their occurrence, glass is of particular concern. As a contaminant its potential to hurt the consumer makes it very emotive - so much so that there have

Chemical analysis of foods

Figure 15 - Typical X-ray microanalysis traces for container glass and Pyrex

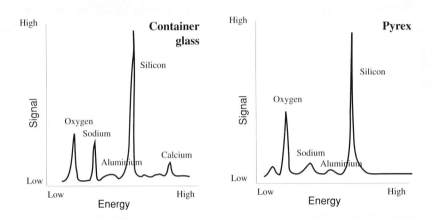

Note that Pyrex contains less sodium (lower sodium peak) than container glass and that container glass contains calcium whereas Pyrex does not.

been highly publicised cases of malicious contamination of products and attempts at extortion. Fortunately such instances are rare, but accidents do happen from time-to-time and if a glass fragment is found in a product it is taken very seriously.

Identifying the type of glass, and hence the likely route of contamination, enables appropriate action to be taken. For example, if the glass has originated in the processing environment the company can take action to withdraw any other affected product, review its glass policy (i.e. its policy on the use of glass in food production areas), and introduce measures to prevent recurrences. Alternatively, if the glass is obviously domestic, as is often the case, then it is likely to have entered the product in the home (e.g. via a chipped bowl).

3.8 Titration

Titration is a controlled chemical reaction from which it is possible to determine the concentration of a test substance (analyte) in an extract. Typically it involves carefully adding to the extract a reagent which reacts with the analyte. When the analyte is all used up, the reagent starts to react with an indicator, causing an easily observed change (e.g. a colour change). At that point, by knowing the amount of reagent added, and by understanding the way the reagent and analyte react together (i.e. how many molecules of one react with each molecule of the other), it is possible to calculate the amount of analyte. Examples of the uses of titration in food analysis are given in Table 12, with cross-references to other sections if they are covered in more detail elsewhere.

Table 12 - Examples of titration in food analysis

Analyte	Comments / purpose
Nitrogen	In Kjeldahl, organic nitrogen is converted to ammonia which is determined by titration with hydrochloric acid (acid-base titration). The amount of nitrogen can then be used to calculate protein and meat content. See Section 3.9.
Ascorbic acid (vitamin C)	Reduces a redox indicator (2,6-dichloroindophenol) to a colourless solution (redox titration). Used for nutritional composition.
Free fatty acids	Titration against alkali (acid-base titration). Used in rancidity testing. See Box 2 p15.
Acidity	Titration against alkali (acid-base titration). Important in preservation (e.g. pickles and sauces).
Chloride	Titration with silver nitrate (Mohr titration). Salt can be calculated from the chloride level (see Section 5.6). Used for nutritional labelling and important in preservation (e.g. pickles and sauces).
Moisture	Karl Fischer method - water is extracted into methanol and titrated against sulphur dioxide, pyridine and iodine.

Different kinds of indicators are used for particular types of titration. Many people are familiar with the idea of pH indicators that are a different colour under acid or alkali conditions. Examples of such acid-base indicators used in titration include methyl orange and phenolphthalein. Another group are the oxidation-reduction indicators - substances which change colour when they are oxidised (lose an electron) or reduced (gain an electron) which can happen in the titration. A third example is the use of starch in iodine titrations as they react to turn the reaction mixture deep blue.

From the examples given (Table 12) it is possible to see that titration is important in ensuring food safety (e.g. checking acidity or salt levels in products which rely on these to prevent microbial growth), for assessing quality and for deriving or verifying information presented on product labels.

3.9 Dumas and Kjeldahl

Analysis of foods for organic nitrogen is an important part of protein and meat content determinations (see Section 5.3). The traditional approach to the determination of nitrogen in foods is Kjeldahl digestion, which is included in various official methods and recognised by international organisations. It tends to provide the baseline against which newer methods such as Dumas combustion - which is also now widely recognised - are compared.

The Kjeldahl method is based on titration (see Section 3.8). The food sample is digested in superheated concentrated sulphuric acid and goes through several processes to convert organic nitrogen in the sample to ammonia, which is then determined by titration (e.g. against hydrochloric acid). In the Dumas method, the sample is combusted at high temperatures (e.g. 900°C) in the presence of oxygen and helium. This produces oxides of nitrogen which are then converted (over a catalyst such as copper) to gaseous nitrogen which can be measured by thermal conductivity.

Both methods have been automated, though the Dumas to a greater extent. Both provide a good indication of nitrogen content, though Dumas values tend to be slightly higher, probably because the method picks up some forms of nitrogen (e.g. nitrite and nitrate, theobromine) that the Kjeldahl does not. The Dumas method is faster (so laboratories can handle more samples in a working day) and, as it does not involve use of reagents such as hot acid, is safer. It also generally uses smaller samples, which can be an advantage, but means that the sample has to be as homogenous as possible so that the result reflects the nitrogen level of the sample as a whole.

Once the level (percentage) of nitrogen in the sample has been determined, it is multiplied by a standard conversion factor to estimate the level of protein in the sample (g/100g). Factors vary slightly from material to material as they are based on the nitrogen content of the proteins present in the type of food - some examples are listed in Table 13.

Table 13 - Examples of factors used for calculating protein content from organic nitrogen content in foods

Food	Factor
Meat	6.26
Milk	6.38
Peanut	5.41
Wheat flour (non-wholemeal)	5.70

To calculate the protein content, the % nitrogen in the sample is multiplied by the appropriate factor. This gives a figure for g protein per 100g sample. For use of nitrogen content in meat content calculations, see Section 5.3

3.10 pH

pH stands for potential Hydrogen. It is a measure of hydrogen ion concentration - that is, of acidity or alkalinity - on a scale of 0-14. On this scale, 7 is neutral, less than 7 is acid and more than 7 is alkali. The scale is logarithmic and inverse, so the higher the hydrogen ion concentration the lower the pH (and vice versa). pH is important as a preservation factor in may products (so called high acid products) as many microbes fail to grow below around pH 4 (see Hutton, 2004 for more on pH in food preservation). It is also important in product quality - for example for getting optimum set of gelling agents in jams and jellies.

The pH of a food is usually measured with a pH meter - this determines the electrical potential (specifically due to hydrogen ions) between a glass electrode and reference electrode using standardised commercially available systems. These usually have temperature compensation systems built into them as the potential is influenced markedly by temperature. To ensure accurate readings, pH meters have to be regularly calibrated using appropriate buffers (solutions of known pH).

For many liquid foods the pH can be measured by dipping the electrode into the sample. For solid or semi-solid materials it is first usually necessary to make a slurry with water (see Table 14). For some moist solid foods (e.g. meat, cheese) it is possible to get electrodes that can be inserted directly into the product. In all cases, duplicate readings are usually taken on the same sample and then averaged.

Table 14 - Preparing foods for pH measurement

Food	Preparation
Liquids (e.g. drinks)	Mix thoroughly by shaking or inversion
Moist foods (e.g. fruit)	Blend thoroughly to uniform consistency with 10-20ml water per 100g product (if required)
Dry goods (e.g. flour)	Mill or blend the sample. Mix 10g with 100ml of water. Incubate at 25°C for 30 minutes with frequent shaking. Allow to settle. Use liquid fraction.
Liquid / solid mixtures (e.g. canned foods)	Homogenise to a uniform consistency and use that slurry or separate into solid and liquid fractions and treat separately as above.

3.11 Water activity

Water activity (a_w) is not to be confused with water content - the two are not the same. It is possible for foods that have a relatively high water content to have low water activity. Jam provides a good example in that it contains plenty of water but has a low water activity. The dissolved solutes in the jam (e.g. sugar) mean that the water is held by the jam and would be unavailable to a microbe trying to grow. Salt has a similar effect on water activity.

Water activity is, as can be seen from this example, extremely important in food preservation. Whilst the water activity of pure water is 1.0, that of a saturated sugar solution is 0.85 and that of a saturated salt solution 0.80; many food poisoning bacteria cannot grow at water activities less than 0.93 (see Hutton, 2004 for more on this).

The water activity of a food is defined as its water vapour pressure relative to that of pure water at the same temperature. Water activity is measured with a hygrometer. A range of these is available commercially. Most involve incubating the food sample in a closed chamber at a particular temperature and measuring the changes in the electrical conductance or impedance of a sensor. This is influenced by the water vapour released by the food as it condenses on a surface being monitored by the sensor. In turn, the release of water from the food depends on the food's water activity. A fuller description of the various systems available and the factors that affect water activity measurement are given in Voysey (1999).

3.12 Immunoassay

Immunoassays exploit the properties of antibodies - proteins produced by the immune system of some animals (hence the term *immuno*assay). Antibodies are produced by animals to help them fend off invading foreign molecules or micro-organisms. Two properties of antibodies make them ideal analytical tools as well as essential components of the animal's defences. First they specifically recognise the molecule / micro-organism that triggered their production. Second, they stick to it tenaciously.

An immunoassay uses antibodies that recognise and stick to the analyte of interest. For example, an antibody against beef protein can be used to detect the presence of beef in a product and will 'ignore' meat from other species, which will not therefore interfere with the reaction. The assay is designed so that the antibody and the analyte can react (i.e. recognise and bind) under controlled conditions (e.g. in a plastic well) and in so doing trigger some measurable change (signal). Commonly this is a colour change but it could also be the agglutination of latex beads, for example. From the change the analyst is able to draw conclusions about the presence and possibly the amount of analyte in the sample.

Some of the applications of immunoassays to food analysis are given in Table 15. An example of one immunoassay format - the sandwich ELISA, in which the analyte becomes sandwiched between two layers of antibody - is shown in Figure 16. There are many variations on the format to suit different types of analyte and different signals, which are described in Jones (2000).

Table 15 - Examples of analytes that can be detected by immunoassay

Analyte / material	Purpose of analysis
Milk and cheese	Determining species of origin for authenticity determination and labelling compliance
Meat, fish and their products	Species testing - to verify authenticity and labelling compliance
Pasta	Checking for non-durum wheat
Gluten (wheat protein)	Verify suitability for coeliac diet (gluten free)
Mycotoxins	Checking against legal limits
Peanut protein	Marker for allergen contamination
Soya protein	Soya content and meat calculations (see Section 5.3)
Antibiotic and other veterinary residues	Contaminant testing / compliance with limits

Figure 16 - The sandwich ELISA (enzyme linked immunosorbent assay)

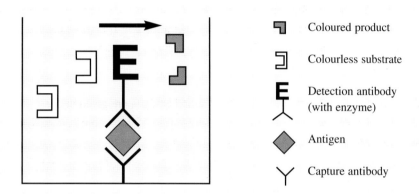

Capture antibody on the test well will capture antigen (if present), which in turn will capture the detection antibody (when added). The detection antibody carries an enzyme which catalyses a colour change. If antigen is present, a sandwich forms and colour develops. If antigen is absent, no sandwich forms and no colour develops. Often the assay can be quantitative (i.e. measure the amount of analyte present) by analysing standard samples containing known amounts of the analyte and relating the resulting colour intensity to the amount of analyte present.

3.13 Electrophoresis

Electrophoresis enables the separation of components in a mixture on the basis of their size and charge. Typically a sample is placed in a small well in a gel (lying on a glass plate) and an electric field is applied across the gel. The components migrate through the gel as they are attracted to the opposite charge of an electrode. Like chromatography there are many variants on the basic theme, and the separation stage has to be followed by some detection system - which usually involves staining the gel to visualise the separated components. Unlike chromatography, electrophoresis is not used particularly widely in routine food analysis, though it does have some important uses. It is mostly applied to the separation of proteins and DNA, and is qualitative rather than quantitative.

The most widely used system for proteins is PAGE (polyacrylamide gel electrophoresis). In a particularly popular version (called SDS-PAGE) the protein is loaded with negatively charged molecules (SDS - sodium dodecylsulphate) before being subjected to the electric field on the gel. The proteins all migrate towards the positive electrode. As they pass through the gel the smaller proteins move more quickly than the larger proteins due to the 'sieving' action of the gel. When stained, the proteins are revealed as a series of bands on the gel. A related but different technique is isoelectric focusing (IEF) in which proteins are separated on the basis of pH.

An example of the use of these techniques is in the identification of fish species. Whilst the species of a whole fish can be identified anatomically (with the appropriate expertise), this is no longer possible once it has been filleted, minced, reformed and blocked. Using PAGE or IEF, each species produces a characteristic banding-pattern, which can be compared with authentic standards to confirm species identity. This is important for authenticity assurance and product labelling.

Another example is with cereal variety identification where PAGE is used to separate the gliadin proteins. This is important because many varieties are better suited to some uses (e.g. bread making) than others (e.g. biscuit production) (see Section 4.8). Certain varieties will command a price premium and variety confirmation will be an important part of a trading agreement. In this case the protein banding-pattern obtained from electrophoresis can be used to identify individual varieties (see Figure 17).

DNA analysis is increasingly being applied to foods and one of the ways that DNA is analysed is by separation of fragments of DNA molecules on the basis of size (see Section 3.14). Traditionally this is achieved by gel electrophoresis, again using charge to 'drag' the fragments through the gel so that the sieving action of the gel sorts them by size. An example of the use of this technique to distinguish between trout and salmon is shown in Figure 18 in Section 3.14, which outlines other examples of DNA analysis, some of which use electrophoresis.

A relatively recent innovation is the development and use of capillary electrophoresis (CE) in which the electrophoresis is carried out in narrow capillary tubes. It can be used not just for the separation of large molecules like proteins and DNA but also for small fragments of proteins and DNA and small organic molecules. Capillary electrophoresis analysis of DNA fragments and/or protein extracts looks promising for a range of applications including cereal variety identification, meat and fish species testing, and detecting GM ingredients. For a more detailed description of this and other electrophoretic techniques, see Wilson and Walker (2000).

Figure 17 - Electrophoretic patterns of different wheat varieties

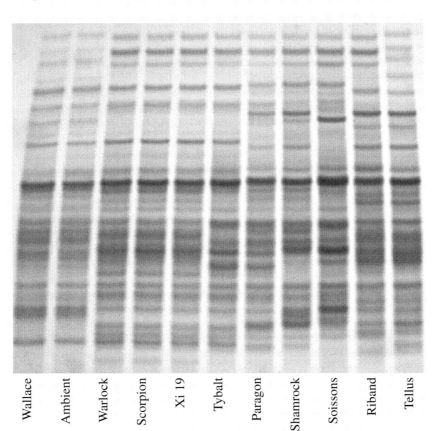

3.14 DNA methods

DNA is the chemical from which genes are made. As food is largely derived from plants, animals and microbes, which all have genes and so all have DNA, most foods contain some DNA. Analysis of this can reveal information about the type and origin of a food material or the ingredients used in a product (see Table 16). Also, because DNA is relatively heat-stable - that is, it survives most cooking processes in a form that can still be analysed - many of the DNA tests applied to food can be used on both raw materials and end-product.

DNA - or deoxyribonucleic acid to give it its full name - is an extremely long molecule which contains information written in a four-letter code. The four letters are A, C, G and T, respectively standing for the names of the four chemicals that project from the backbone of the molecule - adenine, cytosine, guanine and thymine. The order in which these four occur determines the information content of the DNA molecule (like a book written in a four lettered alphabet), which can be read by the cell and, ultimately, translated into activity by the cell's molecular and biochemical machinery.

The important point from the perspective of DNA analysis is that DNA with different sequences of A, C, G and T (i.e. different genes) can be distinguished and/or manipulated by a variety of methods. These include:

- DNA probes, themselves pieces of DNA, which can recognise and stick to specific DNA sequences and be used to reveal their presence - for example, a probe specific to cattle DNA could be used to detect beef in a meat mixture.

- DNA primers - short pieces of DNA that can be used to identify a stretch of 'target' DNA (i.e. a piece of DNA of interest) and trigger its controlled copying (amplification) in a process called PCR (polymerase chain reaction). This results in millions of copies that can more easily be analysed. An example is the detection of genetically modified soya or maize by copying part of the DNA sequence introduced during genetic modification (which is not present in the non-GM crop).

- Restriction enzymes - these are enzymes which cut the DNA chain at specific A,C,G and T sequences, to form fragments which can be separated on the basis of size using electrophoresis (see Section 3.13). The fragment patterns can be compared with each other, with reference materials or with databases of patterns. This is the basis of the original method of genetic fingerprinting and has been applied to verifying the identity of fish species, for example.

Although presented as three alternatives, these approaches can be mixed and matched in various ways to generate extremely powerful techniques. For example, Figure 18 shows how trout and salmon can be distinguished using a method based on amplification followed by restriction enzyme digestion and electrophoresis. Further description of DNA diagnostics in food analysis is beyond the scope of this book but the various methods and their applications are explained in more detail in Jones (2000).

Table 16 - Examples of materials that can be assessed using DNA methods

Material	Purpose of analysis
Meat, fish and their products	Species testing - to verify authenticity or detect adulteration
Meat (e.g. beef)	Sex determination to check supplier specification compliance
Meat / animal tissue	Checking for presence/absence in vegetarian products
GM ingredients (e.g. soya, maize)	Detection for labelling, authenticity and quality control / assurance
Potatoes and rice	Variety identification to check labelling compliance

DNA methods can be used for detecting and/or quantifying a wide range of materials. There are many different DNA-based methods including simple probe tests, fingerprinting, and DNA amplification (using the polymerase chain reaction). In most cases the DNA (which is the analyte being detected) is being used as a marker for the material that contains it. Further details of how these tests work and their applications to food analysis can be found in Jones (2000).

Figure 18 - Distinguishing salmon from trout by DNA analysis

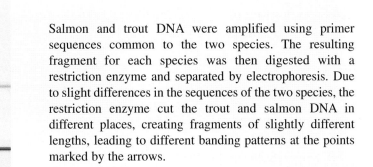

M = DNA size markers
S = Salmon DNA
T = Trout DNA

Salmon and trout DNA were amplified using primer sequences common to the two species. The resulting fragment for each species was then digested with a restriction enzyme and separated by electrophoresis. Due to slight differences in the sequences of the two species, the restriction enzyme cut the trout and salmon DNA in different places, creating fragments of slightly different lengths, leading to different banding patterns at the points marked by the arrows.

4. USES AND EXAMPLES OF FOOD ANALYSIS

Why is food analysed? Who uses the results of analysis? And what actions are taken on the basis of food analyses? This chapter looks at the reasons for food analysis and illustrates these with examples of specific analytes, foods and methods, to demonstrate the importance of getting the analysis right. Many techniques are available to the food analyst, as described in Chapter 3. Each technique will offer particular advantages and will also have its disadvantages. Each will be appropriate for certain analytes and certain food types. In some instances the analyst will have a choice of which technique to use - the approach chosen will depend heavily on the specification of what's required (e.g. qualitative versus quantitative analysis) as well as matters such as cost and turn-around time.

Typical reasons for analysing food products, raw materials and ingredients, include the following and these provide the basis for the examples covered here:

- Determining composition
- Checking for contaminants
- Verifying authenticity and detecting adulteration
- Surveillance and enforcement
- Assessing suitability for purpose
- Supporting product development and product formulation
- Analysis of flavour and flavour changes
- Research into the chemistry and behaviour of food materials

One point to emphasise, as the examples presented in this chapter show, is that these categories are not clear-cut - they overlap considerably. There have been instances of food adulteration, for example, which have led to contamination and serious safety problems. Surveillance and enforcement, although covered in their own right because of their importance, include analysis for composition, contamination and authenticity. Similarly, analysis conducted as part of research projects can highlight new issues of safety, authenticity or raw material suitability, and will generate new methods that might be applied for any of a combination of the above reasons.

So, the categories into which the examples fall are to some extent arbitrary and some crop up in several sections, but they have each been chosen to illustrate a particular area of activity. All the examples illustrate that important decisions are based on analyses, so getting the analyses right is crucial. It is perhaps worth thinking, for each example, what might be the consequences of getting the wrong result in an analysis in the examples described in the following sections.

4.1 Compositional analysis

Analysis of the full or partial composition of a foodstuff is used in various ways including:

- Labelling - analysis can be used to derive nutritional information for the product label or to verify nutritional information derived from standard food compositional tables such as those in McCance and Widdowson (see Food Standards Agency, 2002)

- Routine product safety assurance - for example, the preservation of some foods can depend on threshold levels of salt and pH (e.g. sauces and pickles) or other preservatives such as nitrite (e.g. cured meat) (see Hutton, 2004 for further information) which can be assessed by chemical analysis

- General quality control - for example, determining the quality and authenticity of batches of herbs and spices typically includes a wide range of assessments (see Table 17)

- Checking composition in relation to industry (e.g. supplier-customer) or other specifications. For example, wine exported from Australia to the EU has to comply not only with stringent compositional requirements laid down in the Australian Food Standards Code (to assure the quality of exports) but also additional criteria laid down by the EU and/or individual member states. Furthermore, wine destined for certain retail outlets (e.g. UK supermarkets) has to be accompanied by a certificate of analysis detailing a range of analytes. Taken together, these criteria cover, for example, alcoholic strength, methanol, trace metals, sulphur dioxide, acidity, citric and tartaric acid, and yeast cell wall material (see Stockley and Lloyd-Davies, 2001).

- Product classification for the purposes of excise tariff (see Box 13)

- Government surveys of diet and health - for example, data on composition obtained from analytical surveys (see Section 4.5) can be linked to dietary (questionnaire) surveys to assess intake of nutrients and other dietary components
- Assessment of the safety of novel foods (see Box 14)

The term proximate analysis is sometimes used and usually taken to mean analysis of foods and feedingstuffs for nitrogen (for protein determination), fat, ash (mineral salts), moisture and, by subtraction from the total, carbohydrate.

Table 17 - Typical assessments used to check the quality and authenticity of herbs and spices

Assessment	Method	Purpose
Volatile oil	Distillation and determination by volume	Flavour quality, age of batch, genuineness, and effects of processing
Volatile and flavour compounds	GC and HPLC	Flavour quality
Moisture	Drying or co-distillation of water with toluene	Microbial stability, grinding/flow efficiency
Ash and acid insoluble ash	Incineration and weighing	Detection of bulking material
Fibre	Fibre methods	Detection of bulking materials
Metals (e.g. lead, cadmium, copper, arsenic)	Atomic absorption spectrophotometry	Compliance with legal limits and detection of contamination
Pesticides	GC and HPLC	Compliance with legislation on residues (MRLs) and permitted use of pesticides
Mycotoxins	HPLC and LC/MS	Safety and compliance with legal limits
Specific contaminants or known adulterants	As appropriate	Safety and quality assurance

Box 13 - Chemical analysis of food imports and exports

The movement of goods (including food) between countries is subject to international trade agreements (e.g. the General Agreement on Tariffs and Trade (GATT) and the World Trade Organisation Agreement on Agriculture). Within the EU this ties in with the Common Agricultural Policy and involves systems of export rebates and import duties (see Knight *et al*. 2002 for more on this). Export rebates allow countries to export into markets where they would not otherwise be competitive, whilst import tariffs (duties) prevent markets being flooded with cheaper imports. The system helps to stabilise international trade and is important in all economies. The UK provides a useful illustration.

For imports, tariffs (duties) are payable on many goods. In the UK, import duty totals around £2 billion, around a quarter of which relates to products covered by the Common Agricultural Policy. Internationally, import tariffs can be very contentious as they can be used as barriers to trade. To ensure that import tariffs are applied in a uniform way there is an internationally agreed system for classifying all goods. It is designed such that any one product can be classified in only one group, with rules for allocating goods to categories. In the EU this is embodied in the Combined Nomenclature of the EC. The grouping determines the tariff. Analysis is important in determining whether goods have been correctly classified by companies. There are distinct categories (most with many sub-categories) covering all foodstuffs including meat and meat preparations, dairy products, fish, eggs, vegetables, fruits, nuts, cereals, beverages, and so on.

The import tariff for chocolate provides a good example. In 2004, the duty payable on chocolate slabs of 2 kg or more and containing 18% (by weight) or more of cocoa butter ranged from 8.3% to 18.7% (plus an additional duty for sucrose content). To a chocolate importer, handling large quantities of chocolate, the difference between the lower and upper limits are huge. The point on the scale is carefully determined by the levels of milk fat, milk protein, starch glucose and sucrose / invert sugar / glucose - arrived at by analysis (see Table below). Accurate analysis is therefore very important in determining the correct level of duty that applies.

continued....

Component	Typical methods used
Milk fat	Butyric acid (by gas chromatography)
Milk protein	Derivation from nitrogen content (by Kjeldahl) and lactose content (enzyme test / spectrophotometry)
Starch glucose	Enzyme test (spectrophotometry)
Sucrose / invert sugar / glucose	Enzyme test kit (spectrophotometry)

UK exporters of specific food products - ranging from sugar and cheese to chocolate, biscuits and baked beans - are able to claim rebates from the Rural Payments Agency. The level of rebate payment depends on the nature of the product and/or its ingredients. Again analysis is used in support of this - both for simple products and for more complex, formulated products for which perhaps only one or two ingredients attract a rebate. An example of a simple food that attracts a rebate is sugar (sucrose) - the purity of which is easily determined. In contrast, for a more complex food such as cheese the rebate will depend on the amount of fat in the dry matter (which will involve fat determination by an acid hydrolysis method and moisture determination by weighing following oven air-drying). Similarly, for chocolate the rebate will depend on the level of sugar (determined by enzyme test) and milk fat (determined by the level of butyric acid).

Analysis, then, is an extremely important part of determining the level of rebate for exports and import tariff for imports, and in helping to arbitrate when disputes arise.

References:

Knight, C., Stanley, R. and Jones, L. (2002) Agriculture in the food supply chain: an overview. Key Topics in Food Science and Technology No. 5. CCFRA. ISBN 0-905942-48-5

Anon. (2004) Integrated tariff of the United Kingdom - Schedule of duty and trade statistical descriptions, codes and rates. 2004 edition. TSO. ISSN 0262 0421

Box 14 - Novel food assessment: substantial equivalence

The introduction of a novel food is controlled in the EU by a regulation concerning Novel Foods and Novel Food Ingredients. This defines a novel food as one that has not been used for human consumption to a significant degree in the EU. Before it can be introduced, each novel food has to pass a pre-market approval system: in effect the company wishing to market the food has to submit a dossier of information on the food to the competent authority in the member state in which they first seek approval. In the UK, this body is the Food Standards Agency, which takes the advice of an independent committee of experts - the Advisory Committee on Novel Foods and Processes (ACNFP). The body assesses the dossier and takes expert advice before reaching a judgement on whether the novel food can be marketed. Data on compositional analysis will form an important part of the dossier and be subject to scrutiny - for example, to gauge how closely composition compares with that of similar foods (so called substantial equivalence) or to identify unusual levels of particular components or groups of components.

Two examples of novel foods considered by the FSA in 2004, with contrasting outcomes, are Hawaiian noni juice and Saskatoon berries. Noni juice comes from the noni fruit (*Morinda citrifolia* also known as Indian Mulberry and nonu) which originated in SE Asia but was distributed by ancient voyagers to the Pacific islands, where the juice of the fruit is commonly consumed. A previous application for approval of Tahitian noni juice, produced from the same species of plant but on a different island, had been scrutinised and approved. The composition of Hawaiian noni juice was therefore compared with that of the approved Tahitian noni juice - with analysis of moisture, density, protein, ash, fat, carbohydrate and pH. Allowing for natural variation in the composition, the Hawaiian juice was deemed sufficiently similar to be granted approval.

Saskatoon berries are the fruits of the North American shrub *Amelanchier alnifolia*, grown commercially in Canada since the 1960s and exported to the USA and Japan. Although *Amelanchier* is in the plant family Rosaceae (along with many edible fruits such as apples, pears, plums, cherries and strawberries), visually the saskatoon berry most resembles the blueberry. Although it has a history of consumption in Canada, it would be regarded as a novel food in the

continued....

EU. A company sought approval to market the berries, basing their application on substantial equivalence with the blueberry, as the two are not only visually similar but likely to be used and regarded by the consumer in the same way. The company provided a wide range of analytical data in support of their application, including composition (proximate) analysis, sugar profile, mineral nutrient levels and profile, vitamin levels and profile, fatty acid composition, amino acid composition, and phenolics and anti-oxidants. Despite the wealth of data, the ACNFP decided that drawing the comparison between Saskatoon berries and blueberries - for the purposes of assessing substantial equivalence - was not valid because the two are botanically very different.

References:

Novel Foods and Novel Food Ingredients Regulation EC No. 258/97 of the European Parliament and of the Council.

FSA (2004) Hawaiian noni juice approved (22 July 2004) and Saskatoon berries (20 May 2004). See *www.food.gov.uk* (news archives) for further information.

4.2 Authenticity

An authentic product or food material is one that conforms to its label or the description in any other associated documentation. A product is not authentic if it has been adulterated or substituted. Adulteration is the addition of undeclared material: diluting corn (maize) oil with an inferior vegetable oil but selling it as pure corn oil would be an example of adulteration. Substitution is the replacement of the product with an undeclared alternative: labelling and selling white fish as crabmeat would be an example of substitution. Although this all sounds alarming, surveillance exercises indicate that adulteration or substitution is rare. However, incidences do arise from time to time and chemical analysis plays an important part in authenticity assurance - helping to prevent and detect adulteration and substitution.

Authenticity analysis is used by Government (for surveillance), enforcement authorities (for policing) and industry (as part of supply chain management). It is important that the methods used are reliable - for example, that results obtained by one party in one analysis are consistent with those obtained by a different party doing the same analysis on the same sample (VAM Principle 5). Applying the other principles of VAM will help ensure this.

Where they do arise, authenticity problems are usually the result of deliberate attempts by unscrupulous individuals to defraud others in the supply chain and/or consumers - substituting or bulking-out high value products with cheaper alternatives. Table 18 lists some examples of authenticity problems and the analytical approaches that can be used to help prevent and/or detect them. These methods are used by industry as an integral part of quality assurance programmes (as described in detail in CCFRA 1999) and by industry and government as part of surveillance exercises (see Section 4.5).

With good quality assurance procedures and increasing emphasis on documented traceability of raw materials and ingredients within the food supply chain, accidental adulteration is much less likely. Occasionally, however, adulteration or substitution can happen by accident - by a mix-up over batches or adventitious carry-over. For example, in one incident, traces of undeclared horsemeat found in a salami-type product, manufactured in continental Europe, were almost certainly the result of low levels of cross-contamination during the process. The company, when notified, agreed to change their quality control systems to ensure better segregation of meat species in future - another illustration of action resulting from analysis.

On very rare occasions, adulteration or substitution can make products unsafe, with very serious consequences. Two examples of this are given in Box 15. In the case of Spanish Toxic Oil Syndrome it took some years, and considerable effort by a multidisciplinary team including analytical chemists, to identify the toxic component.

Table 18 - Examples of potential authenticity and adulteration problems and approaches used to combat them

Product	Issue	Analytical solution
Cheese from sheep or goat milk	Dilution of milk with cow milk	Immunoassay
Fish or crustacean	Mixing or substitution with undeclared species	Protein electrophoresis, DNA profiling or immunoassay
Meat and meat products	Mixing or substitution with undeclared species	Immunoassay and DNA analysis
Meat	Selling previously frozen meat as fresh	Enzyme analysis
Vegetable oil (e.g. olive oil, corn oil)	Dilution or substitution with cheaper vegetable oils	Fatty acid profiling by GC and isotopic ratios by MS
Pasta	Bulking out of durum wheat with cheaper non-durum wheat	Immunoassay and HPLC
Basmati rice	Mixing or substitution with inferior varieties	DNA profiling
Poultry	Undeclared irradiation treatment	GC-MS analysis of lipids broken down by the process
Coffee	Undeclared addition of chicory	Inulin analysis and caffeine content
Various	Undeclared GM ingredients (e.g. soya)	DNA analysis
Vegetarian products	Presence of animal material	DNA analysis
Meat and seafood products	Added water	Compositional analysis and calculation

continued....

Product	Issue	Analytical solution
Whisky, vodka and rum	Addition of methanol	GC
Essences and other aromatic ingredients	Mixing / diluting to extend the product	Volatile profiling by GC
Herbs and spices	Bulking to extend the product	Compositional analysis (e.g. moisture, ash)
Milk	Added water	Density, refractive index and freezing point
Wheat grain	Mixing of inferior or inappropriate varieties	Electrophoresis
Wine	Variety and geographical origin	Mineral analysis (e.g. Na, K, Ca, Mg, Mn, Li, Cu, Pb)

All sectors of the food and drink industry face authenticity issues. Industry and Government invest considerable resources in the development of methods to monitor authenticity and detect adulteration of raw materials, ingredients and end products - and thereby prevent problems from arising. The analytical methods listed are typical of the approaches used but these change as new problems emerge and better methods are developed. Also, the analysis itself is but one aspect of a range of procedures and checks that are used to assure food authenticity.

References:

CCFRA (1999) Food authenticity assurance: an introductory guide for the food industry. CCFRA. ISBN 0-905942-21-3

Lees, M. (2003) Food authenticity and traceability. Woodhead Publishing. ISBN 1-85573-526-1

MAFF (1996) Preventing food fraud. Foodsense leaflet PB2533.

Box 15 - When adulteration compromises safety

There are occasions, thankfully rare, when an act of adulteration or substitution is not just fraudulent but jeopardises food safety, sometimes with extremely serious consequences. Perhaps the most serious such incident is that which resulted in Spanish Toxic Oil Syndrome. The incident started as a deliberate act of fraudulent adulteration. A large volume of rapeseed oil had been treated with aniline to downgrade it for industrial use. Some unscrupulous traders decided to refine, decolourise and deodorise this oil, mix it with other oils, package and label it as olive oil, and then illegally introduced it on to the Spanish market (Arribas-Jimeno, 1982; Wood *et al.,* 1994). Unfortunately, the oil contained a highly toxic substance formed in a reaction between the aniline and fatty acids in the oil (Wood *et al.,* 1994).

Over 20,000 people suffered health problems - many with symptoms as severe as respiratory failure and muscle wasting - and as many as 600 people are believed to have died as a result of consuming the oil (Wood *et al.,* 1994). The evidence linking the outbreak to the oil was initially epidemiological, and it was only after several years and extensive investigation, involving sophisticated chemical analysis (including GC-MS) that the causative agent was reliably confirmed (Wood *et al.*, 1994).

Another example is the adulteration of alcoholic drinks with methanol. In 2004 the UK Food Standards Agency issued warnings that counterfeit whisky containing methanol had entered the retail market. Fortunately the FSA was able to issue information about the product (e.g. description of the bottle and label) to enable retailers and consumers to identify it relatively easily - so preventing an incident of methanol poisoning. Others have not been so lucky. The symptoms of methanol poisoning include severe stomach pain, drowsiness, dizziness, blurred vision and blindness and breathing difficulties, and it can be fatal: in March 2003 an Edinburgh woman died after drinking some of an earlier counterfeit brand (FSA, 2003).

References:

Arribas-Jimeno, S. (1982) The Spanish toxic oil syndrome. *Trends in Analytical Chemistry,* **1** (14), IV-Vi.

continued....

FSA (2003) Food Standards Agency issues new warning against drinking counterfeit vodka. Food Standards Agency 25 June 2003 (*www.food.gov.uk/news/pressreleases/2003/jun/vodka250603*)

Wood, G.M., Slack, P.T. *et al.* (1994) Spanish toxic oil syndrome: progress in the identification of suspected toxic components in simulated oils. *Journal of Agricultural and Food Chemistry* **42** (11), 2525-2530.

4.3 Detecting contaminants

In general terms, any chemical that is present in a food when it should not be, could be regarded as a contaminant. However, as this could be true of adulterants and innocuous chemicals, the Food Standards Agency defines contaminants more specifically as 'chemicals whose adventitious presence in foods has the potential to cause toxicological harm to consumers'.

Contamination can arise in different ways and the extent to which it can be avoided very much depends on the nature of the contaminant and the route of contamination. Examples of the main food contaminants, their sources and the foods affected are listed in Table 19. From this it can be seen that the main sources of contaminants fall into several distinct categories:

- The environment
- Production by moulds (mycotoxins) or naturally by the food itself (natural toxicants)
- Formation during cooking / processing
- Migration from packaging or pick-up from other food contact materials

Some contaminants are easier to control than others. For example, dioxins are unwanted by-products of combustion - which includes everything from waste incineration to domestic fires and cigarette smoke. They are persistent environmental pollutants, some of which can harm human health. Their widespread occurrence in the environment as a result of man's activities, means that they find their way into foods: from soil and vegetation they move up the food chain and, being fat soluble, tend to accumulate in the fatty tissues of animals. Foods such as meat, eggs and dairy products tend to be at greater risk. It is widely acknowledged that the best way of controlling dioxin levels in food is through control of emissions. In the meantime, surveillance analysis can be used to ensure that the levels that do occur in foods comply with legal limits and do not compromise food safety to an unacceptable degree. Occasionally, a specific 'acute' problem can arise, and analysis is important in tracking the source of such a problem (Box 16).

Box 16 - A problem with dioxins

During May 1999, news emerged of dioxin contamination of various Belgian poultry and animal products. The problem began in January of that year, when a tank at a fat rendering plant at Deinze in Belgium became contaminated with dioxin, which later that month contaminated a batch of fat used in animal feed manufacture. This in turn contaminated chickens and eggs produced for human consumption. Alerted by problems with the livestock, an investigation was launched. By March it was established that the animal feed was the most likely source of the problem, but the 'causative agent' was unknown. By April/May analyses of the feed, livestock and eggs confirmed dioxin contamination as the cause of the problem. From records kept by the rendering plant and others in the supply chain, it was possible to identify all the animal feed producers that had been affected: eight in Belgium, one in France and one in the Netherlands. From here it was possible to identify 417 poultry producers that had bought affected feed: the products of these companies were withdrawn.

The European Commission issued two decisions, the first dealing with the feedstuffs, fowl and poultry products, and the second extending the restrictions to cover Belgian pigs, cattle and derived products. The UK MAFF liaised closely with industry to trace and withdraw any potentially affected products and to offer advice (via a consumer helpline). Throughout, analysis and supply chain records were used together to help identify, assess and control the problem. Setting aside the accusations that the Belgian authorities were slow to inform European consumers of the problem, it is certainly clear that the incident would have taken considerably longer to resolve without the power of chemical analysis.

Reference:

For a fuller description of the incident see Hall, M.N. and Jones, J.L. (1999) Public confidence in fertilisers and in food quality. International Fertiliser Society Conference Proceedings No. 441 'Quantifying responsible plant nutrition'.

Table 19 - Examples of potential food contaminants and their sources

Contaminant	Source	Affected foods
Dioxins	Environment - unwanted by-products of combustion (e.g. incineration, bonfires, cigarette smoke)	All foods but mainly animal products - e.g. meat, milk, butter, eggs, fish - due to accumulation.
PCBs (polychlorinated biphenyls)	Environment - manufactured between 1930s and 1970s	Mainly animal products - e.g. meat, milk, butter, eggs, fish - due to accumulation.
PAHs (polycyclic aromatic hydrocarbons)	Environment - combustion and refining of fossil fuels	Mainly animal products and vegetable oils.
Metals	Environment - natural occurrence, mining, combustion and industrial activities	Various including crops, animal products and fish
Tin	Cans (unlaquered)	Occasionally elevated levels in some canned tomatoes and canned fruit
N-nitrosamines	Formed during processing	Alcoholic beverages, fermented foods and cured meats
Chloropropanols (MCPD and DCP)	Formed during processing	Soy sauce, hydrolysed vegetable protein, some baked goods, cooked/cured fish
Ethyl carbamate	Formed during processing	Fermented products especially alcoholic drinks
Marine toxins	Algae	Shellfish (accumulated from algae by filter feeding)
Mycotoxins (e.g. aflatoxins, ochratoxin, vomitoxin)	Formed by certain moulds	Cereals, nuts, dried fruit, coffee
Patulin (a mycotoxin)	Formed by certain moulds	Apples and apple products
Acrylamide	Formed during processing	Fried and baked starchy foods including potato products, bread and savoury snacks
Sudan dyes	Added as an adulterant to 'improve' colour	Some chilli powders and products containing these powders
Veterinary residues (e.g. antibiotics, growth promoters)	Animal husbandry (if not used in accordance with good agricultural practice)	Meat, dairy, fish and their products

continued....

Contaminant	Source	Affected foods
Pesticide residues	Crop husbandry (if pesticides are not used in accordance with good agricultural practice)	Crop products
Semicarbazide	Seals / gaskets on lids of glass jars	Various products sold in jars with such lids
BADGE (Bisphenol A diglycidyl ether)	Migration from can coatings	Canned fish, soup, meat, fruits, vegetables, desserts, beverages, pasta
Cleaning chemicals	Residues from cleaning operations	Various processed products

The table lists examples of chemicals or groups of chemicals that can contaminate food, and the sources of such contaminants. Although the list might appear long and daunting, it is important to bear in mind that successive surveillance exercises show that serious contamination is rare. As described in the text, analysis plays a very important part in bringing such incidents under control when they do arise. It is also important in routine surveillance of 'at risk' foods (see Section 4.5).

By way of contrast, the formation of mycotoxins in some foods is often easier to control. Mycotoxins are produced by some species of moulds that can contaminate foodstuffs such as cereals, nuts, dried fruit, coffee, and spices. Examples include the aflatoxins, ochratoxins, and trichothecenes. Taking cereal grain as an example, mould can only grow on the grain if it becomes sufficiently damp and warm. Controlling the humidity and temperature of the grain store can limit mould growth and prevent the formation of mycotoxins. Because mycotoxins are acutely or chronically toxic or carcinogenic, products susceptible to mycotoxin contamination are routinely surveyed to monitor the levels (see Section 4.5).

Examples of toxic compounds that can form during processing include chloropropanols and acrylamide. Chloropropanols form in various foods but particularly soy sauce and the savoury ingredient hydrolysed vegetable protein. The main chloropropanols of concern are 3-MCPD (3-monochloropropane-1,2-diol) and its derivative 1,3-DCP (1,3-dichloropropanol). Because 3-MCPD and 1,3-DCP can cause cancer in laboratory animals when they are fed large amounts throughout their lives, there are concerns that they may present a risk to human health, even though they are consumed at much lower levels. Following their discovery in foods, limits were introduced and measures adopted to minimise their levels in foods. Analytical surveys were then conducted to ensure compliance and reassure consumers that any risk they did pose was minimised. When surveys like this are conducted to monitor or compare the levels of specific dietary components over time, obtaining reliable data is essential if the comparison is to be valid. This illustrates the importance of the philosophy embodied in the principles of VAM.

The discovery that acrylamide can form in foods such as fried potato products during processing has fuelled a great deal of research. This is because acrylamide is a toxin and is known to cause cancer in animals (though it is currently not known whether acrylamide in the diet presents a significant hazard to human health). Research, including methods for the analysis of products for acrylamide, will help in understanding the problem and devising solutions (see Section 4.10).

Semicarbazide (SEM) provides an example of a contaminant arising from a food contact material - in this case packaging. In July 2003 it became apparent that semicarbazide occurs as a chemical by-product in the plastic seals of metal twist-off lids of glass jars. There is very little evidence of its effects on humans but some studies suggest it is a weak carcinogen in animals. The levels in food appear to be very low. Although the risk to human health is also thought to be extremely low, the UK Food Standards Agency and European food bodies commissioned research to determine the levels in food as well as other measures such as looking for alternative seals in which SEM does not form. Obviously these lines of research will require methods for detecting and quantifying SEM.

Another area where food could become chemically contaminated through food contact materials is in picking up traces of cleaning fluid from processing equipment which has not been rinsed sufficiently. This can some times lead to taint problems, and chemical analysis is used in identifying and helping to solve such problems (see Section 4.7).

In all the examples listed here, analysis plays an extremely important part in identifying sources and monitoring the levels of specific contaminants - and so is an essential part of food safety assurance. The number of analytes involved is enormous, so it is important that the analyst and those commissioning the analysis are clear about what is required (VAM Principle 1). Mercury, for example, can occur in two forms in food - free mercury and organic mercury. The latter is more toxic. Depending on the levels found, an analysis that determines each type could be more meaningful than one that estimates total mercury.

Finally, one perceived problem that arises from chemical analysis - and particularly from the ongoing development of methods that are able to detect ever-lower amounts of compounds - is the discovery of 'new' contaminants. Acrylamide is a good example: being a product of a long-used process, acrylamide is likely to have been part of our diet for a very long time. It is also only because of increasingly powerful analytical techniques that we can become aware of such problems and introduce measures to combat them.

4.4 Pesticide and veterinary residue detection

Just about every crop grown is open to attack from pests and diseases and to competition from weeds. Pesticides are used to protect crops from such damage - by preventing the attack before it starts or, more usually, limiting its spread once it starts. Different types of pesticides fulfil different roles (Table 20). In each case, the pesticide is used to kill, or at least prevent the growth of, the target organism. Because pesticides are designed to kill they have to be used with great care. Plant growth regulators, which are used to control the growth or cropping pattern of crops, are also included within the term pesticide, though strictly speaking they are not pesticides as they do not kill pests.

There are strict controls on which pesticides can be used with which crops, the amounts that can be used, the timing of applications (particularly in relation to harvesting), and the levels of residues on the final crop. A residue is any pesticide that remains on the harvested product and maximum residue levels (MRLs) are prescribed by law.

The MRL is the legal maximum amount of pesticide that is allowed to remain on an agricultural commodity (e.g. fruit, vegetable) after its use in accordance with good

Table 20 - Examples of pesticides and their targets

Type of pesticide	Used to control:
Insecticide	Insects and diseases transmitted by insects (e.g. many plant viruses)
Fungicide	Fungi - moulds, mildews and rots
Herbicide	Weeds (i.e. plants other than the crop)
Nematocide	Nematode pests

agricultural practice. It is an offence to exceed this limit. MRLs have been set for a wide range of pesticides on a wide range of products. They reflect the maximum amount of pesticide that would be expected if it had been applied in accordance with the terms of its regulatory approval (e.g. time and amount of application, period between application and harvest [harvest interval], crop type) and good agricultural practice.

Analysis is used to verify compliance with MRLs. It can also be used to check that unauthorised pesticides have not been used. It is also used as part of wider surveillance exercises. In the UK, these are overseen by an independent committee (the Pesticide Residues Committee) which advises government. As the analysis is being used in support of legislation and trading agreements, it is important that results obtained in different locations are consistent (VAM Principle 5) if costly disputes are to be avoided. Also, with so many compounds it is important that the analyst and the user of the analysis are clear about the requirements of the analysis (VAM Principle 1) before it is undertaken.

Typically, analysis for pesticides is based on gas chromatography with a suitable detection system such as flame photometry, electron capture detection or, in more sophisticated laboratories, mass spectrometry (see Section 3.3 and 3.4). Methods exist for screening for groups of pesticides: for example, dithiocarbamate fungicides can be analysed collectively by reducing them to carbon disulphide (CS_2) and detecting this by GC-flame photometry. Organophosphorus (OPs) and organochlorine (OCs) insecticides can each be analysed as groups - using GC-flame photometry or GC-MS (for OPs) and GC-electron capture or GC-MS (for OCs). Some compounds cannot be analysed as part of a group and require 'targeted' analysis: a good example is the plant growth regulator chlormequat. This is used, for example, to limit stem extension in cereals and reduce lodging (where the stem bends or breaks). Analysis is by LC-MS-MS.

In some instances, the MRL might be set at the limit of detection for that pesticide: in other words, there should be no detectable amount of that pesticide present. This generally arises in one of three circumstances:

- Where the use of the pesticide on a particular crop or in particular circumstances is not supported in the EU
- Where scientific data show that the intended use might leave residues that would pose an unacceptable risk
- Where scientific data show that the intended use of the pesticide leaves no determinable residues on the treated commodity at harvest

Veterinary residues - arising from the use of veterinary compounds in animal husbandry - are analogous to pesticide residues in many ways. Veterinary medicines are used to treat or prevent illness, and particularly infectious disease, amongst livestock. For example, antibiotics are used to control bacterial infections, other drugs to treat infections with parasitic worms or fungi, and insecticides to control fleas. Vaccines might be used to prevent the spread of infections. Table 21 lists examples of some of the veterinary compounds used.

Before their use is authorised, veterinary medicines have to fulfil three criteria: quality (i.e. that they can be produced to an appropriate and consistent standard), efficacy (i.e. that they do the job it is claimed they do) and safety (for consumer, handler and target animal). In some instances veterinary products will leave residues (e.g. in milk, meat, eggs) either of the medicine itself or of products formed by breakdown of the medicine. As with pesticides, there are provisions to control the levels of such residues, including specified MRLs. Some veterinary compounds

Table 21 - Examples of veterinary compounds and their role

Example	Function
Antibiotics*	Treatment of infection with micro-organisms (e.g. cow with mastitis)
	Promoting growth
Anti-parasitic treatments such as coccidiostats and acaricides	Kill or prevent the growth of parasites (e.g. coccidiosis in poultry) including mites and ticks (acaricides)
Vaccines	Prevent infections becoming established and spreading
Anti-fungals	Treat fungal infections (e.g. of skin, hooves)
Hormones and beta-agonists*	Increase growth rate, increase leanness

*Use of hormones and beta-agonists to promote growth is not permitted within the EU but is in some other regions (e.g. US). Use of antibiotics to promote growth to be phased out in EU by 2006.

are used for growth promotion, to increase the growth rate or lean meat content. However, although the use of hormones to increase the growth of livestock is permitted in some regions (e.g. US) it is not permitted within the EU.

Analysis is important in both of these circumstances - that is, checking for compliance with MRLs or ensuring that prohibited compounds are not being used. Within the UK, regular surveillance of UK-produced animal products (e.g. meat, milk, poultry, eggs, farmed fish) and a wide range of imported processed products for veterinary medicines is overseen by the independent Veterinary Residues Committee which advises government. More widely, surveillance is also carried out by the FAO/WHO Joint Expert Committee on Food Additives. In addition, companies will monitor suppliers. A range of methods is used to detect veterinary residues, reflecting their chemical heterogeneity. They include inhibition of microbiological growth (for antimicrobials), immunoassay (for individual residues or groups of related residues), HPLC with UV, fluorescence or MS detection, and GC with electron capture, infrared or MS detection (see Chapter 3).

Organic food is another area where the use of both pesticides and veterinary compounds is restricted and, again, where analysis can be used in support of compliance. Veterinary medicines can, in fact, be used in the treatment of animals intended for production of organic food - and particularly in the treatment of disorders and diseases that would cause suffering. However, in such cases, products from the animal would either not be marketed as organic or would be subject to a specified withdrawal period between administration of the medicine and production of the organic foodstuff. Although the use of pesticides is also restricted in the production of organic crops, several pesticides are permitted. These include potassium permanganate (as a fungicide and bactericide), sulphur (as a fungicide and acaricide), calcium polysulphide (as an insecticide, fungicide and acaricide) and paraffin oil (insecticide and acaricide). Again, analysis is an important part of monitoring compliance and of surveillance exercises.

4.5 Surveillance exercises

Analytical surveillance exercises can be carried out locally (e.g. by a local authority), nationally (e.g. by a government agency), supra-nationally (e.g. by the EU), and across a supply chain (e.g. by a retailer, food manufacturer or representative body). Such surveys may be conducted in response to a specific incident or problem or carried out as part of a long-term structured programme.

For example, in the UK the Food Standards Agency (FSA) carries out many food surveys (see Table 22). These are mostly as part of an ongoing programme of routine monitoring of product safety and authenticity as part of the FSA's remit for public protection, but *ad hoc* surveys are also conducted in response to specific incidents or findings. The routine programmes tend to focus on particular products known to be at risk from specific hazards (e.g. certain cereals, nuts, dried fruits and coffee are susceptible to mycotoxin contamination) or open to potential adulteration by unscrupulous operators (e.g. adulteration of corn oil with cheaper oils). They might also look at particularly sensitive issues - for example where meat might be adulterated with that from a species (e.g. pig) which is not consumed by some groups on the basis of their religion.

Other surveys are aimed at building a picture of dietary intake of nutrients or of exposure to specific chemicals through the diet: analytical surveys can be used with surveys of eating habits (or specific diary surveys) to estimate intakes of a wide range of chemicals. These estimates can then be used to compare actual (or likely) intakes with desirable intakes or safety guidelines (i.e. sufficient nutrients and minimal contaminants in relation to risk assessment and safety thresholds). A good example of the latter is the particular attention paid to infant foods in surveillance of dioxins and PCBs.

The *ad hoc* surveys are usually commissioned in response to surveys elsewhere, new research or perhaps just general suspicion. For example, in 2003 reports emerged from Australia that dichloropropanol (DCP) had been found in meat. By this time it was widely known that DCP occurs in soy, for example, but it had not been found in meat products (see Section 4.3). The UK FSA instigated a survey of a wide range of meat products and was able to establish that DCP was not a problem in UK meats.

Sometimes surveys are instigated on the basis of rumour rather than official reports. For example, in response to anecdotal evidence that certain salami type products imported into the UK from continental Europe contained undeclared horsemeat, the FSA instigated a pilot study and follow-up survey. The use of horsemeat in these products is not prohibited so long as it is declared, but the FSA felt that consumers would wish to be aware of which, if any, of these products did contain horsemeat. The survey of 158 samples from 30 regions across the UK confirmed that there was not a problem: only 1 sample was found to contain traces of horsemeat, at the limit of detection, as a result of cross-contamination.

Often regulatory agencies work together. For example, local authorities will often collect samples on behalf of the FSA, providing coverage across the country as a whole or targeted to specific areas as appropriate. The FSA itself will conduct

Table 22 - Examples of FSA surveillance exercises

Analyte or issue	Food	Purpose	Method
Salt, fat and sugar	Pizza and sausages	Nutritional composition in relation to dietary advice	Standard compositional analysis
Colours	Sweets and soft drinks	Compliance with legal limits	HPLC
Horse meat	Salami	Authenticity	DNA analysis
Varietal identification	Potato	Authenticity - accurate varietal labelling	DNA analysis
Fatty acids and other components	Maize (corn) oil	Authenticity (presence of other vegetable oils)	GC and carbon isotope ratios
Varietal identification	Basmati rice	Authenticity	DNA analysis
Chloropropanols	Soy sauce, meat	Safety	GC-MS
Ethyl carbamate	Whisky	Safety	GC-MS
Metals	Various foods	Dietary intake	ICP-MS
Mycotoxins (aflatoxins and ochratoxin)	Nuts	Safety	HPLC and LC-MS
Mycotoxins (tricothecenes)	Oat products	Safety	GC-MS
Mycotoxins	Rice and rice products	Safety	HPLC and GC-MS
Sulphur dioxide	Soft drinks	Compliance with legal limits	Titration
Dioxins	Various foods	Dietary intake for safety assessment	GC-MS
Brominated flame retardants	Trout and eels	Dietary intake in relation to safety	GC and LC-MS

At the time that this table was compiled (2004) the FSA had also published its plans for surveys to be included in its future programme of activities. These included: mycotoxins in baby and infant foods; aflatoxins and ochratoxin in spices; aflatoxins in nuts; mycotoxins in liver and kidney; nitrate in spinach and lettuce; MCPD, acrylamide, and nitrosamines in various foods; metals in allotment produce; organic tin in shellfish; arsenic in fish and shellfish; dioxins and PCBs in fish and shellfish; and sulphur dioxide in various products.

Many of the analytes listed form the basis of proficiency testing schemes (see Section 2.9) which provides a means for laboratories to participate in independent assessment of their technical performance (VAM Principle 4).

Source: Food Standards Agency (*www.food.gov.uk*)

surveys as part of larger EU-wide programmes. For example the EU has in place a programme for monitoring the levels of dioxins and PCBs in the diet, to assess intake and any safety implications. In the UK the survey is co-ordinated by the Food Standards Agency, which forwards its results to the European Commission to collate and assess along with results from similar programmes conducted in other member states.

Surveys within industry - of raw materials, ingredients or end products from suppliers, for example - are often confidential, forming part of a company's quality control processes. Some programmes are routinely reported however. For example, in the US, milk in all bulk tankers is checked for veterinary residues, such as antibiotics, and any testing positive are kept out of the food supply chain. The test results are submitted to the National Milk Drug Residue Database (developed jointly by the Food and Drug Administration and National Conference of Interstate Milk Shipments) and published in a report. In 2003, of over four million samples tested from across all 50 states, less than 3,000 (less than 0.07%) were found to contain veterinary residues.

The previous sections on authenticity and contamination, and the attention paid to these by government and enforcement bodies through surveillance, could create the impression that problems are widespread. The results of the surveys, however, generally suggest the exact opposite. The vast majority of samples analysed in surveys conform to legal requirements, labelling declarations or specified safety limits. Furthermore, where problems are identified action plans are devised and implemented to address the problem and monitor progress. Ethyl carbamate in whisky provides a good example of this.

Ethyl carbamate is a chemical formed naturally during fermentation. It has been shown to be a carcinogen in animals and is regarded as a potential carcinogen in humans (though, when consumed, most is excreted or degraded in the body within 24 hours). Surveys of a range of fermented drinks and foods, conducted by the UK Ministry of Agriculture, Fisheries and Food in the 1990s, indicated that the largest intakes of ethyl carbamate were via whisky. Studies on the formation of ethyl carbamate - and particularly the role of raw material composition and specific processing variables - led to recommendations that could reduce ethyl carbamate levels. A follow-up survey published by the Food Standards Agency in 2000 found that the levels of ethyl carbamate in whisky were considerably lower as a result of the adoption of these recommendations by the industry.

A similar story holds for MCPD (a chloropropanol) in soy sauce. As described above (Section 4.3), this compound forms in various foods during processing, with the highest levels found in soy sauces. Modifications to the process can significantly reduce the levels formed. Successive survey exercises show that the proportion of samples containing levels of MCPD above the legally permitted level is falling, following identification of the problem, recommendations for avoiding problems and the introduction of legal limits.

The examples covered in this section demonstrate the importance that analysis plays in supporting the monitoring of food safety and authenticity and illustrates the role that analysis plays in decision making in government and industry. It again illustrates the importance of consistency between analytical measurements in different locations (VAM Principle 5) and, implicitly, of the measures that need to be adopted to help ensure this (see Chapter 2).

4.6 Legislation

Many of the examples of uses of food analysis covered in this chapter relate to legislation. This is not surprising as legislation plays an important part in protecting consumers (in terms of safety and fraud), providing definitions of many terms used in product descriptions and promotions, and setting standards to which food producers, manufacturers, retailers and caterers have to adhere. Examples of how chemical analysis is used to support and enforce legislation, and already covered in this chapter, include:

- Labelling and product descriptions - ensuring that the product conforms with the information presented on the pack (e.g. product name, nutritional information, ingredient declarations and specific claims such as 'low fat')
- Compliance with maximum residue levels for pesticides or veterinary residues

- Ensuring that contaminants such as mycotoxins or heavy metals do not exceed legal permitted levels
- Verifying whether specific foodstuffs do or do not attract entitlement to import duty or export subsidy

A few other examples further illustrate the relationship between legislation and chemical analysis of food and drink. The UK Food Safety Act 1990 is important in upholding food safety in the UK, by providing a 'catch-all'. Amongst its many provisions it states that it is *"an offence to sell to the purchaser's prejudice any food which is not of the….substance ….demanded by the purchaser."* Coeliacs have an intolerance to gluten (a proteinaceous component of wheat) and have to avoid foods that contain it. Some foods are marketed as 'gluten free' on the basis that they contain alternatives to wheat. A product which is labelled 'gluten free' but which turns out not to be so would not be of the 'substance demanded' by a coeliac purchaser acting on the information on the pack. It would also be to the 'purchaser's prejudice' if they consume it and it makes them ill. Chemical analysis is used by companies that make such foods to ensure that the product is gluten free (i.e. of the substance demanded). It is also used by enforcement bodies to investigate where problems arise.

In addition to this type of 'catch-all' legislation, other regulations provide criteria in relation to specific safety issues. Erucic acid is a fatty acid found in some varieties of rapeseed and mustard seed. High levels of erucic acid have been linked with the build up of fatty deposits in the heart muscle of animals. Although there have been no cases of erucic acid poisoning in humans, the varieties of rapeseed and mustard seed used for food have been bred so that they contain very low levels of erucic acid. There are also UK regulations (The Erucic Acid in Food Regulations 1977) which limit the erucic acid content of foods to no more than 5% of the total fatty acid content of foods that contain more than 5% fat. Chemical analysis is used to support this regulation, including periodic surveillance of 'at risk' products.

Food additives provide another example. Not only do various pieces of legislation control what additives can be used in what foods and at what levels, but EU legislation also lays down purity criteria for a number of additives. Analysis is used to support, uphold and refine these criteria. For example, ethylene oxide is used in the production of polysorbates - which are used as emulsifiers in bakery, dessert and confectionery products. Some ethylene oxide inevitably carries through into the additive as an impurity, but there are limits of 0.5-1.0 mg/kg for the level of residue in the additive. Following analytical improvements, with reductions in the limit of detection to 0.2 mg/kg, the EU's Scientific Committee on Food (in an opinion expressed on 17 April 2002) recommended that the limits should also be reduced to 0.2 mg/kg.

Where analysis is used in support of legislation - which can result in prosecutions and other legal judgements - reliable analysis is paramount. The assurance of reliable analysis through the sorts of measures outlined in Chapter 2 will also help to ensure consistency of measurements in different locations (VAM Principle 5) which is also particularly important where legal judgements are involved.

4.7 Flavour, off-flavour and taint analysis

Flavour is an important component of food quality and derives from the food's volatile compounds (i.e. aroma) and its taste. Off-flavours are generally defined as unpleasant odours or tastes resulting from the natural deterioration of the food. Taints are usually regarded as unpleasant odours or tastes resulting from contamination of a food by one or more extraneous chemicals. Understanding flavour, off-flavours and taints - their formation and the way they can change during processing and storage - is important in understanding product quality, in process control and product development, and in troubleshooting.

The ultimate judgement on flavour quality, or on the presence of off-flavours or a taint, will rest with consumers' sensory reaction to the product. On a more sophisticated and controlled level, objective sensory analysis is an extremely important part of industry's approach to assessing flavour and quality. However, chemical analysis is important too - indeed, sensory and chemical approaches are often used in parallel. Troubleshooting a taint problem provides a good example: a consumer complaint of taint might initially be assessed by sensory analysis, which can help describe the taint objectively and give the analytical chemist a good idea of which chemicals to target in the analysis. A definitive chemical identification, by say gas chromatography or mass spectrometry, can then give clues as to where the taint chemical might have come from, or how it formed.

Taints can arise from many sources including packaging materials, cleaning fluids (e.g. carry-over due to poor rinsing), strong smelling materials which come into contact with the product (e.g. through inappropriate storage) and even microbial contamination (though the latter is often regarded as off-flavour). Sometimes taints might arise through a combination of these. For example, a major pan-European project was undertaken in the mid-1990s to explore the causes of musty taint in wine and to reduce its occurrence. The taint is largely due to 2,4,6-trichloroanisole (TCA), produced by commonly occurring moulds before and during cork production and transportation. High levels of TCA can also be associated with the occurrence of a rare but specific cork defect called yellow stain. However, although this can contribute to musty taint in wine, such taints are not always due to cork. Also TCA is a common cause of taint in other food and drink products.

Wheat provides another example where the off-odour is generated not by the food itself but by microbial contaminants. If the moisture content and temperature of the grain store are too high, then a combination of fungal and bacterial activity can generate off-odours - in particular musty, sour, 'green' and fishy odours. Interestingly, these odours in wheat are regarded as off-flavour but the problem with cork, described above, is regarded as taint. This illustrates that grey areas exist, especially where terms are applied before the origin of the problem is known (as is the case with cork taint). What is clear is that chemical analysis can be very important in helping to unravel complex problems such as these.

As well as using chemical methods to explore the causes of taint problems, new methods are constantly being sought to help with troubleshooting through the rapid identification of low levels of taint compounds. In a recent project at CCFRA, a GC-MS method was developed for the detection of chlorotoluenes at low levels in a simple food matrix (milk). These compounds, which occur in cleaning fluids, have distinctive mass spectra and look promising as markers for this potential source of taints.

Chemical methods are also important in exploring the vast range of volatile compounds that give foods their characteristic odour: in almost all cases the characteristic odour of foods arises because of the mix of volatiles it releases, rather than any single compound. For example, the typical odour of a fruit can result from the combination of a couple of hundred different volatiles in varying proportions - and the pattern will change as the fruit ripens. Sometimes the patterns can be used as markers - for example for authenticity testing or distinguishing between different varieties of the same fruit (see Box 17). In meat, to take another example, the volatile profile of fresh and rancid material will be very different (see Box 2 p15), making a very different assault on the nose.

Box 17 - Apple flavour

The characteristic flavour and aroma of fruits is largely due to the release of volatile organic compounds. The production and release of perhaps dozens or more compounds increases during ripening, accounting for the build-up of the full-ripe flavour. Different fruits give off different combinations of volatile compounds and in different amounts - giving them a characteristic 'volatile profile' - and even varieties of the same fruit can differ in their volatile profile. These profiles can be assessed by gas chromatography, with attendant practical applications.

For example, certain volatiles produced by apples - namely hexanoic acid, 2-hexenal and alpha-farnesene - could be used to detect the presence of apple puree in strawberry puree (Reid *et al.* 2004). This is a useful approach to authenticity testing where adulterators might be tempted to bulk-out expensive purees (such as strawberry) with cheaper purees (such as apple). In this case the volatiles are not unique to apple but statistical analysis of their relative amounts, determined by gas chromatography, can be related back to apple to act as a 'marker' for the adulteration. In a separate study, a similar rapid method, again based on the statistical analysis of a volatile profile, was developed for distinguishing between new apple cultivars. This was done as a system with which to develop and assess the method, rather than for direct application to apple analysis, but in the long term it has potential applications in fruit quality assessment and in early, objective assessment of fruit aroma in plant breeding programmes.

References:

Reid, L.M., O'Donnell, C.P. and Downey, G. (2004) Potential of SPME-GC and chemometrics to detect adulteration of soft fruit purees. Journal of Agricultural and Food Chemistry **52**(3) 421-427.

Schulz, I., Ulrich, D. and Fischer, C. (2003) Rapid differentiation of new apple cultivars by headspace solid-phase microextraction in combination with chemometrical data processing. Food **47**(2) 136-139

4.8 Checking suitability for purpose

Analysis can form an important part of checking that a particular material is suitable for the purpose for which it will be used. A simple example might be a company manufacturing a gluten-free product for coeliacs, who are gluten intolerant. Immunoassay test kits (see Section 3.12) can be used as part of quality assurance and quality control to make sure that the raw materials conform with the company's specification as regards gluten.

A broader example is provided by wheat, as grain from different varieties can have very different properties. The quality of wheat supplied to the milling industry is monitored to ensure that each consignment of flour despatched to the baking industry has the properties which make it suitable for specific end uses - including bread, biscuits, pasta products, batters, sauces and pastry production. For example, flour used to make bread must have good quality protein that produces gluten of adequate strength (see Box 18).

When wheat is delivered to a grain store or mill it may be subject to a number of quality tests, which can include the following chemical analyses:

- The moisture content of the grain will be measured - this is important because high moisture can lead to premature sprouting of the grain, encourage the growth of moulds with the possible formation of mycotoxins, and encourage the growth of micro-organisms that lead to off-flavours and odours. In the laboratory, moisture will typically be determined by weight loss on oven drying (see Section 3.2).

- Tests on protein content and quality will be conducted. The visco-elastic properties of the gluten are vital for most end uses. A particular problem is gluten damage when the grain is dried at excessively high temperatures before storage. Protein content should attain about 13% for breadmaking. Typically protein content is determined by Dumas combustion (see Section 3.9).

- The variety of wheat may be verified by an electrophoresis test (see Section 3.13) - this is important because different varieties are suited to different uses (e.g. bread, biscuits).

- The level of *alpha*-amylase enzyme in the grain is an important indicator of quality. Amylase digests starch and is formed naturally in the grain during sprouting. High levels of *alpha*-amylase will indicate that the grain has started sprouting during storage and that it is unsuitable for many baking applications. Amylase activity is influenced by wheat variety and by wet weather at harvest. Although traditionally determined by measuring the viscosity of a gelatinised flour-water suspension (known as a 'Falling number test'), amylase activity can also be measured biochemically, by measuring the formation of a coloured compound in an amylase mediated conversion of a test substrate (by spectrophotometry - see Section 3.6).

Box 18 - Protein content of flour

Gluten is a proteinaceous material found in wheat dough and formed from proteins found in flour. The quality of the gluten in a batch of wheat is critical in determining suitability for breadmaking. It is well established that within a given breadmaking wheat variety the higher the protein content the greater the loaf volume - and this is where gluten plays its role. During bread production the yeast ferments sugars within the dough to carbon dioxide, which becomes trapped in the dough matrix to give the bread its characteristic bubble structure. Good breadmaking flours produce a gluten which has a relatively high resistance to deformation, high extensibility and moderate elasticity. This helps preserve the bubble structure created during mixing and allows significant expansion during proving (when the dough is left to rise) and baking.

The protein content and gluten quality is influenced not only by the type of wheat used (e.g. some varieties are better suited to breadmaking while others are better for products such as biscuits) but also by agronomic factors. Application of nitrogen fertilisers, for example, can improve the protein content of flour from breadmaking wheat.

Further reading:

Cauvain, S. and Young, L (2001) Baking problems solved. Woodhead Publishing Ltd.

Bhandari, D. (2000) The early prediction of breadmaking quality of grain and its improvement through targeted late application of nitrogen fertiliser. HGCA Report No. 219.

4.9 Product development

Product development is an extremely important part of the success of food and drinks companies. New products are needed to help companies remain competitive and to move into new markets. Thousands of new products are launched every year in the UK alone (see Hutton 2001 for more on this). Product development is sometimes seen as an entirely creative process - where marketing people 'brainstorm concepts' and home economists throw together new combinations of ingredients to try out new product formulations or completely new ideas.

However, a large part of product development involves rigorous and logical structured testing of alternatives and some requires sophisticated chemical analysis. A couple of examples illustrate where chemical analysis can fit in product development.

The first involves efforts to reduce the trans fatty acid content of confectionery (Kristott, 2004). Many products contain fats and oils as ingredients. The physical properties of these (e.g. melting point) can be manipulated by blending different fats and oils. For example, those with longer chains and which are saturated tend to be solid fats and lards. Those with shorter chains or which are unsaturated fatty acids tend to be lighter oils. Blending these can alter the properties.

The properties can also be altered by chemical hydrogenation. This introduces more hydrogen into the fat and so increases the level of saturated fats. However, this process can also change some of the remaining unsaturated fats - so that some of the more common cis form is changed to the trans form. Trans fats have been linked with heart disease in recent years and, although the relationship is not clear-cut, governments and industry have moved to reduce the levels of trans fats in the diet. Chocolate itself contains very low levels of trans fats. However, some of the fillings used in filled chocolate confectionery can contain fats which are perhaps 50% trans fats. Various ways are being devised of developing new products with similar appeal but much lower levels of trans fats (see Kristott, 2004). An important part of this process, however, is the rigorous analysis of the alternatives for their relative levels of cis, trans and saturated fats.

Another example - again related to diet and health - centres on antioxidants. Several degenerative diseases - including cancer, stroke and cardiovascular disease - are thought to be exacerbated by oxidative stress, that is oxidation reactions within particular tissues in the body. The corollary of this is that consumption of foods rich in antioxidants, such as anthocyanins and polyphenolics, could help prevent or retard the progress of such degenerative conditions. Certainly epidemiological evidence suggests a link between high fruit and vegetable consumption and reduced risk of these diseases. Consequently, a range of new and existing products are being promoted on the basis that they are rich in antioxidants. Chemical analysis has been used to profile and assess particular product formulations such as mixtures of fruits and fruit juices and the effects of various processes on their antioxidant capacity (Chaovanalikit and Wrolstad, 2004). For example, HPLC can be used to determine the relative levels of specific compounds and ORAC (oxygen radical absorbance capacity) to measure the capacity of the product to absorb oxygen free radicals.

4.10 Analysis as a research tool

Food analysis is an important part of research as it is used to develop better understanding of food materials, the effects on foods of food processes, interactions between products and their packaging, and various aspects of food safety. For example, the discovery that acrylamide forms in starch-rich foods at high temperature (see Section 4.3) resulted from research in Sweden in 2000. This, in turn, spawned a wave of research projects, across the globe, in which analysis of foods for acrylamide was used to assess the types of products and processes most affected: an important first step in trying to reduce the levels of acrylamide and improve product safety. The same is true for compounds such as 3-MCPD in soy sauce and hydrolysed vegetable protein and semicarbazide migration from plastic seals into product (see Section 4.3). Sometimes specific incidents require extensive research to resolve the problem - Spanish Toxic Oil Syndrome is perhaps a good example of this (Box 15).

But the use of chemical analysis in food research is not just about troubleshooting. Examples of research projects in which chemical analysis plays an important part include:

- Monitoring the nutritional quality (e.g. vitamin content) of foods processed or stored in different ways

- Understanding the chemical basis of raw material performance
- Tracking the fate of agrochemicals, used in crop husbandry, during onward processing, storage and preparation of food products
- Understanding the effects of fertiliser application on protein content and breadmaking quality of wheat
- Assessing the factors that affect the levels of natural toxicants or contaminants in particular products
- Looking at the effect of livestock diet on the composition of meat, milk and eggs - from the viewpoints of nutrition and processing
- Determination of levels of biomarkers in studies of diet and health (see Box 19)

In all these examples, analysis is important in helping us understand the food we eat and the factors that affect its safety and quality. Projects such as these, and therefore chemical analysis by definition, contribute hugely to assuring the supply of safe and wholesome food. And this research is all in addition to the vast amount of research into the development and improvement of the methods of chemical analysis themselves.

Box 19 - Analysis of biomarkers in studies of diet and health

One of the difficulties of carrying out studies of the health effects of specific dietary components, is assessing how much of that dietary component a person actually consumes. Analysis of biomarkers can help with this. Biomarkers are chemicals that can be used to provide a reliable indication of a person's intake of some material.

For example, it has been suggested that there is an association between high intake of caffeine by pregnant women and low weight of their child at birth. In the UK, the Food Standards Agency launched a project to explore this. The idea was to monitor the caffeine intake of a large group (3,000) of pregnant women, record the weight of their children at birth, and assess whether there was a relationship between the two. One of the classic problems with such studies is reliably monitoring the intake: in this case intake of caffeine-containing beverages. Humans being what they are, some of the participants will forget to record their intake. Others might under-record it, feeling that they should be

continued....

seen to reduce their intake. And the beverages themselves will vary in caffeine content. And what about caffeine from sources other than beverages? By conducting chemical analysis of biomarkers - in this case caffeine and caffeine metabolites - in urine and saliva, at specific points in the pregnancy, and combining this with questionnaires and intake diaries, a much more reliable picture of caffeine intake can be constructed.

A contrasting example - in this case of a foodstuff believed to confer health benefits - is provided by whole grain cereals. Consumption of these has been linked to a decrease in the risk of obesity, diabetes, cardiovascular disease and some cancers. It is believed that these protective effects result in part from phenolic compounds in the bran of whole grain. One of the major groups of phenolics are the alkylresorcinols (ARs). As well as offering potential health benefits in their own right, these compounds offer potential as biomarkers for cereal whole grain consumption as they can easily be detected using methods such as HPLC and GC. Similarly there is considerable interest in the health benefits of flax. Particular attention has focused on alpha-linolenic acid (ALA), a polyunsaturated omega-3 fatty acid (see Hutton 2002 for more on this), and lignans (another group of phenolic compounds). Again, metabolites of lignans have been used as biomarkers of flax-lignan intake, in studies of the health benefits of flax.

In each case, the analysis of biomarkers can give a far more reliable indication of intake of the 'parent material' than could a diary or survey alone, so that analysis adds significantly to the quality of information from such studies. This, in turn, can lead to better decisions and more effective use of the study findings.

References:

FSA (2003) Agency study into possible link between caffeine consumption and low birthweight. Food Standards Agency News, March 2003, p3.

Oomah, D. (2004) Perspective on flax based on clinical studies. Food Science & Technology **18**(2) 40-42.

Ross, A.B., Kamal-Eldin, A. and Aman, P. (2004) Dietary alkylresorcinols: absorption, bioactivities, and possible use as biomarkers of whole-grain wheat- and rye-rich foods. Nutrition Reviews **62**(3) 81-95.

5. COMPLICATIONS AND COMPROMISES IN FOOD ANALYSIS

Some analyses sound like they should be quite simple - perhaps because we are familiar with what they represent in everyday terms - but are very complicated. For example, information we take for granted on the label of a pack can be complicated to derive through analysis: meat content, dietary fibre, energy content, vitamins and salt all provide good examples of where analytical complications can arise and/or compromises have to be made.

In some instances the analyte cannot be measured directly because it is not a single chemical but a complex mixture. Dietary fibre, for example, is a mixture of complex molecules that together form indigestible matter. Meat is largely a variable mixture of proteins, fats and water. In instances like this, the definition of the analyte (e.g. fibre content, meat content) goes hand-in-hand with the analysis: the definition and the analytical approach depend on each other.

In other cases, chemicals we group together because of their function often include compounds with very different chemistries. For example, vitamins are grouped together because of their nutritional role not their chemistry. Although collectively defined by their function 'as organic substances essential to human life in very small amounts' (see Hutton 2002 for more on this), in chemical terms they are extremely varied and their analyses have to be approached in different ways.

Then there are examples where the compound of interest has to be determined via one of its components. Salt provides a good example of this: much of the salt (sodium chloride) in food dissociates into its constituent sodium and chloride ions and has to be determined via analysis of either or both of these.

This chapter looks at some of these examples in outline, to illustrate that despite having a rigorous scientific basis, chemical analysis is also something of an art, where the experience of the analyst and, occasionally, compromise have to be used to circumvent problems that might otherwise be insurmountable.

5.1 Dietary fibre: an empirical approach

Dietary fibre is the indigestible matter in food. More precise definitions vary (for more on this see Kirk and Sawyer 1991, and Hutton 2002). One thing that is agreed is that fibre is a complex mixture of molecules, some of which are themselves very complex. This reflects the intricate chemical structure of the plant cell walls from which dietary fibre is derived. The way in which dietary fibre is defined, therefore, depends on the chosen method of analysis. This is called an empirical approach: fibre is difficult to define in chemical terms, therefore it is defined in terms of what can be measured. Two contrasting methods illustrate the point:

- The AOAC (Association of Official Analytical Chemists) method is based on a broad definition of fibre including complex polysaccharides (such as pectins, gums, mucilages, cellulose and resistant starch [starch that can escape digestion in the gut]) as well as non-polysaccharide material such as lignin. The method involves removing protein and non-resistant starch by enzymic digestion, weighing the residue and then deducting from this the ash (inorganic mineral salts - determined separately) and any residual protein (also determined separately).

- The Englyst method, by way of contrast, concentrates on non-starch polysaccharides (NSP). Starch is removed from the sample by enzymic digestion before the other polysaccharides are broken down into their component sugars (acid-hydrolysed) which are determined colorimetrically. The standard procedure does not include resistant starch, but modifications exist to allow its inclusion.

In the UK, in the absence of a legal definition of dietary fibre, the Food Standards Agency recommends the AOAC approach as the preferred option for determining fibre for nutrition labelling. What is more important, however, is that the method used is made clear to anyone acting on the information so that they are aware of the definition to which they are working. For example, the UK has a system of dietary reference values (DRV) for various nutrients. They give a guide to the amount of the nutrient needed each day to maintain good health. For fibre, the DRV of 18 g of non-starch polysaccharide equates directly to an Englyst determination. However, if measured by the AOAC method, then a daily intake of 24 g of dietary fibre can be used as a guideline.

5.2 Global migration: an empircal approach

Another example where the parameter is defined in terms of how it can be measured is global migration. When certain products and packages are held in close contact, material can move from the package into the product. This is called migration. Under EU and UK regulations, food packaging components must not be transferred to food during its normal shelf-life if this is to the detriment of the food (e.g. if it poses a health risk to the consumer or adversely affects product quality) (for more on this see Hutton 2003).

Migration of specific compounds can be monitored in the same way that any analyte can be measured: through extraction followed by analysis with techniques such as chromatography. However, this will only pick up those compounds specifically sought. So an alternative 'catch-all' approach can also be used - to determine total or global migration. In this approach the packaging is weighed and immersed in a standard matrix (e.g. olive oil is used to simulate a fatty matrix) under defined conditions, then the oil removed and the packaging reweighed. However, there is a complication because the package will absorb some oil that cannot be removed simply by drying. A correction has to be made for this: the oil is extracted with a solvent and then quantified by gas chromatography; this is possible because the oil is a relatively simple and well defined matrix. Global migration is calculated as the original mass of the pack less the mass of the pack after the migration less the mass of the absorbed oil. A similar approach can be taken with aqueous simulants to assess global migration into water-based foods.

This empirical approach allows global migration to be defined in terms of the approach used and allows direct comparisons to be made between different packaging materials, or the same material under modified but controlled conditions (e.g. at different temperatures). It is an approach that could not be used with more complex food matrices as it would not be practicable to quantify the amount of matrix retained by the packaging during the test.

5.3 From nitrogen to protein and meat content

Meat is not a single analyte but a complex mixture of chemicals including protein, water and fat. The meat content of a product is best estimated at the mixing bowl stage - in terms of the amount used as an ingredient. However, in practice things are not necessarily this simple. For example, for quality control or enforcement purposes it is often necessary to calculate the meat content based on analysis of the end product. Also, there are legal definitions on what can and cannot be regarded as meat (e.g. the amount of fat and the amount of connective tissue) and meat content declarations have to reflect this. It is also possible for some of the meat in a product to 'disappear' during processing (e.g. meat fat can migrate into pastry in a pie). Calculations of meat content, based on analysis, have to take account of issues such as these.

The amount of meat in a product can be calculated from the nitrogen content, with allowance made for nitrogen derived from other sources (e.g. vegetable or cereal protein which doesn't contribute to the meat) and for the usual fat content of the meat (which does). A full description of meat content calculations is given in McLean (1999). The example presented here illustrates the basic approach. But first, a couple of definitions are needed:

- *Apparent total meat content* is an estimate of the meat content of a product derived from calculations based on data obtained from product analysis
- *Apparent fat-free meat content* is the calculated meat content with no allowance for any fat that is present in the product and originating from the meat

The starting point for a meat content calculation is a compositional analysis of the product, including determination of nitrogen content and fat. Over the years, tables of data have been generated relating the apparent fat-free meat content to the nitrogen content for particular meat species and cuts. Examples of these factors are given in

Table 23. By determining the nitrogen content of a product (see Section 3.9) the analyst can use the appropriate nitrogen factor to calculate the apparent fat-free meat content using the following equation:

$$\text{Apparent fat-free meat content (\%)} = \frac{\text{\% total nitrogen}}{\text{Nitrogen factor}} \times 100$$

So, taking as an example a beef sausage found on analysis to contain 1.90% nitrogen, and knowing that beef has a nitrogen factor of 3.65, for this sausage:

$$\text{Apparent fat-free meat content} = (1.90/3.65) \times 100 = 52.1\%$$

However, this does not allow for the fat derived from the meat. Suppose the analysis of the sausage also indicated that it contained 19.1% fat. The apparent total meat content is calculated by adding this to allow for the fat:

$$\text{Apparent Total Meat Content} = 52.1 + 19.1 = 71.2\%$$

This is a relatively straightforward case from which it is possible to see how an estimate of meat content can be derived from an analysis of nitrogen and fat. In reality it can become much more complicated. For example, allowance might have to be made for non-meat nitrogen (e.g. from soya, cereal or casein) which could increase the apparent meat content by pushing up the nitrogen levels. Such ingredients can be determined by other methods (e.g. immunoassay, microscopy) so that they can be taken into account. Also there will be limits on the amount of fat that can be included as meat - so it might not be permissible to attribute all the fat in the product to the meat - and different methods of fat analysis can generate different results (see Section 5.5). It is even possible to end up with more than 100% meat (see Box 20).

Table 23 - Examples of nitrogen factors for meat content calculations

Material	Nitrogen factor
Pork (general)	3.50
Beef	3.65
Lamb	3.50
Chicken (whole)	3.70
Turkey breast	3.90

The nitrogen factors are determined experimentally by analysing known authentic meat samples (e.g. species, meat cuts, no added water) and relating nitrogen content to known amount of meat. For a fuller list of nitrogen factors used in the calculation of meat content and a full explanation of the calculations see McLean (1999).

Box 20 - Not less than 120% meat?

The basis of the meat content calculation can cause some confusion - not least when a product label declares that the product contains more than 100% meat. This often causes amusement in correspondence columns of newspapers and magazines, and prompted an explanatory letter to New Scientist drawing attention to the clarity of labelling on a Hungarian salami (New Scientist, 17 June 2004, p31). The label declared "Pork (min. 161%); fat, salt, spices, preservative E260. Made with 161 grams of raw pork meat per 100 grams of salami. Moisture is lost during curing and maturation".

The distinction is between a statement of the ingredients used and the composition of the product on analysis: the two are not the same. This is why the apparent total meat content of a simple piece of roast meat can be around 140% - because moisture has been lost during cooking. In effect, the analysis of meat content is used to determine the amount of meat that was present at the recipe (mixing bowl) stage - but this is, at best, an estimation.

Box 21 - Added water in scampi tails?

Scampi - also known as Dublin bay prawn and Norwegian lobster - is the tailmeat of the crustacean *Nephrops norvegicus*. Some scampi is sold raw but much is sold coated (e.g. breaded) and flash fried. Much water is used during the processing of shellfish to achieve good manufacturing and hygienic practice (e.g. to remove shell). However, this can complicate scampi content calculations. For example, during washing the shellfish tend to absorb water and lose soluble protein, which affects the nitrogen levels and so the apparent scampi content. Also, the levels of nitrogen in fish and shellfish vary enormously anyway, not just with species but with size, sex, spawning cycle and fishing ground. A further complication for scampi is that it can be ice-glazed (i.e. sprayed with water and frozen) to protect it against dehydration and oxidation during storage.

To accommodate all this, a joint industry/enforcement code of practice was developed, which gives guidance on good manufacturing practice, making allowance for some inevitable absorption of water and loss of protein, levels of glaze and so on. The code also includes agreed nitrogen factors to calculate shellfish content and added water.

Towards the end of 2001, the FSA organised a survey in which a number of local authorities from across the country collected samples of shellfish, including ice-glazed scampi and breaded scampi, from fishmongers, supermarkets and catering suppliers. Compositional analysis was carried out on these to enable determination of percentage scampi content using the standard 'meat content' calculation as follows:

$$\%\text{scampi} = \frac{\%\text{total nitrogen} - \%\text{non-meat nitrogen}}{\text{Nitrogen factor}} \times 100$$

For the ice glazed scampi, the percentage added water was determined by measuring the total amount of moisture (% moisture) and subtracting from this the amount of water naturally associated with scampi. The amount of 'natural water' is calculated as the amount of scampi (%scampi) less the amount of protein (%protein) - as scampi tail is mostly water and protein (the contribution made by carbohydrate and ash is negligible). The percentage added water is therefore given by:

continued....

Complications and compromises

$$\% \text{ added water} = \%\text{moisture} - (\%\text{scampi} - \%\text{protein})$$

One particular precaution that had to be taken with the ice-glazed scampi was to check that appreciable amounts of soluble protein had not moved into the ice-glaze during storage (as this would reduce the nitrogen in the scampi and affect the result).

What this illustrates is that the determination of parameters such as fish content or added water draws upon a range of analyses and involves various assumptions (e.g. the relationship between nitrogen content and 'meat' or 'fish' content). It also requires careful thought to avoid unforeseen errors (e.g. accidental nitrogen loss).

As for the survey, of the breaded scampi samples, around a quarter were found not to display the scampi content declaration required by law or declared up to 5% more scampi than was present. A further quarter contained up to 22% more scampi than was declared. The added water content of the ice-glazed scampi ranged from 9-44%. Manufacturers of products with inadequate declarations of added water or scampi content were notified and local authorities asked to follow up the results and check compliance with the agreed code of practice.

References:

FSA (2002) Survey of added water in raw scallops, ice-glazed (peeled) scampi tails, and scampi content in coated (breaded) scampi products. Food Survey Information Sheet 30/02.

Anonymous (1998) Code of practice on the declaration of fish content in fish products. Joint publication of UKAFFP, BFFF, BRC, BHA, Seafish Industry Authority, LACOTS and APA. March 1998.

The calculations can also be extended to determine the level of connective tissue (e.g. rind, skin). This is done by measuring the amount of hydroxyproline (an amino acid found in collagen and which provides a marker for connective tissue) and making an allowance for this (with due regard to specified legal limits). Typically hydroxyproline would be determined colorimetrically (see Section 3.6). Another extension to the calculation is to determine the level of added water (see Box 21). Usually added water is determined by calculation, with subtraction from 100% of other known components (e.g. apparent total meat content, salt, carbohydrate, and other ingredients).

Even in outline it is evident that the determination of meat content for a composite product requires bringing together data from a number of analyses and offers various complications. Similar principles, with similar complications, apply to the calculation of 'meat' content in fish and crustacean products, though for some of these the absorption of water during processing can cause an additional complication (see Box 21).

5.4 Energy content

Food provides us with two things: raw materials and energy. Various biochemical reactions within our body enable us to harness the energy in a form that it then uses to fuel growth, development and maintenance. The standard (SI) unit of energy is the joule. The amount of energy in our food can be measured in joules or, more meaningfully, in kilojoules (kJ - thousands of joules). On the label of a product the energy content is usually declared both in kilojoules and, because it is more familiar to many consumers, calories. The two can be inter-converted on the basis that 1 calorie is approximately the same as 4.18 joules (and 1 kcal, which is confusingly often written as 1 Cal, the same as 4.18 kJ).

The energy in food can either be directly measured or it can be calculated using standard procedures based on its composition - and the two approaches won't necessarily provide the same answer. This is because not all of the energy in foods is available to the body - for example, a lot of energy in the food is tied up in the indigestible matter (fibre) and lost in the faeces - and so is excluded from energy calculations based on composition. Even the energy from molecules that are digested and absorbed into the body is not necessarily used or retained by the body - for example, some of the energy from protein in the diet is lost through urea excreted in the urine.

The total chemical energy in a food (called the gross energy) can be measured in a device called a bomb calorimeter - this measures the amount of heat released from the food during complete combustion. Typically this value will be around 10% higher than that derived by calculation based on standard energy-conversion factors applied to the food's components. Under UK food labelling regulations, the energy declaration on the product label should be derived by calculation as this better reflects the energy available to the body than does the measurement of gross energy - i.e. it excludes energy lost with indigestible material and so on. Examples of conversion factors for calculating energy content are given in Table 24.

Energy content determination is another good example of why it is important to be clear on what information is required and to what use it will be put when deciding which approach to take (VAM principle 1).

Table 24 - Energy content by calculation

Component	Conversion factor (kJ/g)
Carbohydrate	17
Protein	17
Fat	37
Alcohol	29
Organic acids	13
Polyols (e.g. sorbitol)	10

Following compositional analysis, the amount of each material (in grams of the material per 100 grams food) is multiplied by the conversion factor to determine its contribution to the energy content per 100 grams of food. To convert to kcal (Cal) each factor should be divided by 4.18. Note that some components contain much more energy than others, and so contribute more 'Calories' to the food. For example, gram for gram, fat contains over twice the energy of carbohydrate and protein, and alcohol nearly twice the energy of carbohydrate and protein.

5.5 Fat content

Fat content provides another example where the analysis sounds simpler than it is in practice. As is evident from earlier sections, accurately determining fat content is important not only in its own right (e.g. for nutritional information declarations), but also affects other calculations (e.g. meat content, energy content). Different approaches generate different results, which might also be influenced by the food type, so it is important for the user to understand what the method is actually measuring and how it relates to the information required.

The fat present in food is not a single compound but a mixture of different compounds which can include fatty acids (with various chain lengths and degrees of saturation), triglycerides and phospholipids, for example. Some might be free (i.e. not attached to anything) whereas others might be bound (attached) to proteins or other food components. Free fats are generally easily extracted from food using solvents such as chloroform or petroleum spirit. However, foods such as meat, which contain bound lipids, will require treatment (hydrolysis) with acid or alkali in order to release the bound fat. Analysis without the hydrolysis will generally suggest a lower level of fat than analysis with hydrolysis, as it will not include the bound fat. So, if data on free fat instead of total fat were used in a meat content calculation, it would lead to an underestimate of the meat content (see Section 5.3). Similarly, if it were used in an energy calculation (see Section 5.4), it would lead to an underestimate of the energy content.

A range of methods of fat analysis has been developed, including gravimetric methods (extracting the fat and weighing it), volumetric determination (extracting the fat and measuring the volume), gas chromatography, and calculation by difference. The latter is generally used for butter, and involves measuring moisture and salt and then deducting these from the total mass to estimate the fat - though more direct methods for measuring fat in butter are under development.

Of the various gravimetric methods two examples illustrate the distinction between analysing free fat and total fat (i.e. free plus bound fat). The Bligh and Dyer method (for free fat in a range of products) involves macerating the food in a methanol-chloroform mixture followed by further addition of chloroform and water to generate two phases with the fat dissolved in the chloroform phase. This is separated and the solvent evaporated to leave the fat, which can be weighed and/or used in further analyses. In the Weibull-Stold method, the sample is boiled in dilute hydrochloric acid to digest any protein and liberate free fat, and then dried before the fat is extracted with solvent. Although this approach gives total fat, the sample cannot always be further analysed for fatty acid profile (by GC) because of damage to fatty acids during acid hydrolysis.

The decision on which method to use is often a matter of experience, and should take into account whether the need is for free fat or total fat and the possibility of further analysis. For example, to determine the total fat content of a chocolate biscuit with a nut and cream centre, the method would need to include a hydrolysis step to release the bound fat in the nut that will otherwise go undetected. In some instances specific methods might be stipulated in legislation or otherwise recommended by an official body. For example, in the EU the Rose-Gottlieb method is the official method for the determining fat content of milk powder and condensed milk, while the Weibull-Stold method is specified under the UK Animal Feedingstuffs Regulations 1999 for total oil content of animal feed. McLean and Drake (2002) have reviewed the main methods for analysis of fats and oils in foods, including more detailed consideration of the points raised in outline here.

5.6 Sodium versus salt

Analysis for salt has been touched upon in earlier sections (e.g. see Chapter 1 p6 and Chapter 2 p13-14). Salt is sodium chloride. In solution, including the aqueous phase of foods, it dissociates into its two components - sodium ions and chloride ions. The amount of salt in a product is derived by calculation following analysis of the product for sodium (e.g. by flame photometry/atomic absorption - see Section 3.5) or chloride (e.g. by titration - see Section 3.8). Each gram of salt contains approximately 0.39g sodium and 0.61g chloride. So, the salt content can be calculated by determining the amount of sodium and multiplying this by 2.54 (i.e. by 1/0.39) or by determining the amount of chloride and multiplying this by 1.65 (i.e. 1/0.61).

Box 22 - The wrong approach, the wrong result?

As emphasised in Chapters 1 and 2, being clear about the purpose of the analysis and the way in which the result will be used is just the first step in the whole analytical approach. Even when the most appropriate method is chosen, the range of quality assurance measures described in Chapter 2 should be deployed wherever possible, if the analyst and user are to have full confidence in the result. However, if an inappropriate method is used, no amount of quality assurance will rectify the situation. The table below pulls together some of the examples from this section, to illustrate the effects of using an inappropriate method. For example, for some products an analysis of free rather than total fat will lead to an underestimate of fat content, which could affect the legality of nutritional and meat content declarations.

Approach	Result
Analyse for free fat rather than total fat	Underestimate of fat content, energy content and meat content
Energy determination by bomb calorimeter rather than calculation	Overestimate of energy content
Salt via sodium	Overestimate of salt levels if the food contains sodium from non-salt sources
Salt via chloride	Overestimate of salt if the food contains chloride from non-salt sources. Underestimate of salt if chloride is lost during ashing
Fibre via Englyst or AOAC	Care in comparing with dietary reference values (18 g for Englyst and 24 g for AOAC)

However, foods often contain other sources of sodium (e.g. a common raising agent is sodium bicarbonate, curing can involve sodium nitrite, monosodium glutamate is used to enhance flavour). In such cases, relying on sodium content will lead to an overestimate of the salt levels. Similarly, some foods contain chloride from non-salt sources (e.g. from potassium chloride); in such cases relying on chloride content could again lead to an over-estimate of the salt level.

The compromise is to choose whichever is more appropriate for the particular product (bearing in mind non-salt sources of sodium and chloride) and the use to which the data will be put. For example, for a nutritional declaration on a product label, sodium might be what is needed, irrespective of whether it derives from NaCl or non-NaCl sources, so a figure based on sodium analysis would be more appropriate.

5.7 Vitamins are not one group

The vitamins are chemically a heterogeneous group. Some, such as vitamins A, D, E and K are fat-soluble whilst vitamin C and the B vitamins are water-soluble. Beyond this, the sub-groups are quite complex in some cases. Vitamin A activity, for example, is mostly due to retinol, which can exist freely or can be linked to fatty acids as retinyl esters, and which is available only from animal sources (e.g. liver, milk). However, the carotenoid group of fatty plant pigments also provide an important dietary source of vitamin A indirectly: beta-carotene, for example, can be cleaved to yield two molecules of the vitamin.

The B vitamins are even more complex - thrown together by virtue of their function. The original vitamin B activity was designated as something that combated the illness beri-beri. The group includes thiamin (B1), riboflavin (B2), niacin (nicotinic acid and nicotinamide), pyridoxal derivatives (B6), folates (a range of molecules derived from folic acid), pantothenic acid, and biotin. By way of contrast, vitamin C (ascorbic acid) is one compound. A description of the roles of the different groups and their occurrence and use in foods (e.g. fortification, anti-oxidants), is given by Hutton (2002).

Table 25 - Examples of the methods used for vitamin determination in food

Vitamin	Method(s)
Vitamin A	HPLC with UV absorption for detection; spectrophotometry
Vitamin B - folate	Micro-bioassay
Vitamin B - niacin	Micro-bioassay; colorimetry
Vitamin B - pantothenic acid	Micro-bioassay
Vitamin B - riboflavin	Micro-bioassay
Vitamin B - thiamin	Micro-bioassay
Vitamin C (ascorbic acid)	HPLC with UV absorption for detection; indophenol titration
Vitamin D3	HPLC with electrochemical detection
Vitamin E	HPLC with electrochemical detection; TLC with spectrophotometry; colorimetry

The analysis of foods for vitamins involves a range of approaches (see Table 25). Some can be determined using HPLC linked to an appropriate detector - for example UV absorption for those that absorb light or electrochemical detection where an oxidation-reduction reaction gives a better signal. In the case of the B vitamins, however, the chemical heterogeneity and lack of methods for individual compounds means that they are often determined in terms of activity by means of a bioassay involving an assessment of microbiological growth.

6. CONCLUSIONS

This book set out to explain some basic principles of food chemical analysis and, by way of examples, to put them in the context of the day-to-day activities of the users of food analysis - analysts, industry, government and enforcement authorities. The wide-ranging examples presented, whilst not comprehensive, illustrate how food is analysed, why it is analysed, and why it is important that food is analysed properly.

There are, overall, three main themes:

- *Food analysis is important.* Analysis is conducted to generate information and this is then used to make decisions - often big decisions. The wrong decision could prove extremely costly. It could, for example, jeopardise product safety and affect consumer health and wellbeing, or it could adversely affect business performance by damaging brand image and triggering the needless withdrawal and destruction of valuable stock.

- *There are schemes, procedures and systems that analysts can follow to help ensure that their results are reliable.* These include, for example, staff training programmes, laboratory maintenance and calibration regimes, method validation procedures, accreditation schemes and proficiency testing schemes - that is, the kind of activities that are required to implement the six principles of VAM:

    1. Analytical measurements should be made to satisfy an agreed requirement
    2. Analytical measurements should be made using methods and equipment that are fit for purpose
    3. Staff making analytical measurements should be competent to undertake the task
    4. Technical performance of a laboratory should be regularly and independently assessed

5. Analytical measurements made in one location should be consistent with those made elsewhere
6. Organisations making analytical measurements should have well defined quality control and quality assurance procedures

- *Proper implementation of these principles and procedures adds to the cost of analyses, but it is a cost worth paying.* The laboratory that takes care to implement these systems will incur costs in doing so, which means it will have to charge more for its services than will a laboratory that does not follow such procedures. However, the cost brings a benefit in terms of the reliability of the results generated, and is likely to be negligible compared to the cost of getting the analysis, and hence the decision, wrong.

New methods are under development all the time, and existing methods are subject to on-going refinement. In taking an overview, this book provides a basis for more specialised reading on the subject. The references and websites listed in the final section should provide a useful starting place for this.

7. GLOSSARY AND COMMON ACRONYMS

The correct use of terminology is important if misunderstandings are to be avoided. This glossary gives definitions of some common terms in chemical analysis, with particular emphasis on those that cause confusion or are often used incorrectly (e.g. 'precision' and 'accuracy'). As acronyms can also cause confusion, some of the more common ones are also included here. Where appropriate, terms discussed elsewhere are cross-referenced. They are also explained in more detail in the VAM booklet '*Introduction to measurement terminology*' and in some of the other books on laboratory quality management cited in the references section.

Accreditation - formal recognition by an authoritative body that a laboratory or person is competent to carry out specific tasks. A laboratory's accreditation will have a defined scope, which specifies which activities have been accredited. This means that laboratories are often accredited for particular methods but perhaps not for others. Examples include accreditation by UKAS (the United Kingdom Accreditation Service) and the Campden Laboratory Accreditation Scheme (CLAS). For more information see Section 2.8.

Accuracy - the closeness of the result of a measurement (i.e. analytical result) to the true value (i.e. the actual amount of substance present). Note that accuracy is different from precision and the terms should not be used interchangeably. For results to be accurate they must be precise and unbiased. See below and Box 7 p35.

Analysis - the measurement of the amount of a substance (the analyte) in a sample, or the detection of the presence of an analyte in a sample, using an analytical method. Everything in the sample apart form the analyte is the matrix.

Analyte - the substance (or group of substances) for which the food or other material is being analysed. Everything in the sample apart from the analyte is the **matrix**.

Bias - the difference between the average of a number of measurement results and the true value (the actual amount of the analyte present). See Box 7 p35.

FAPAS - food analysis performance assessment scheme - a UK based proficiency scheme (see below and Section 2.9).

GC and GLC - gas chromatography and gas-liquid chromatography. See Section 3.3.

HPLC - high performance (or pressure) liquid chromatography. See page 54.

Limit of detection - the smallest amount of a substance that can be reliably detected in a sample (without necessarily determining the amount present).

Limit of determination / quantification - the smallest amount of a substance that can be reliably quantified in a sample.

Measurement uncertainty - provides an estimate of how 'certain' the analyst is about the result of a measurement. Uncertainty is expressed as a range of values within which the true value of the quantity being measured is expected to lie.

MS - mass spectrometry. Identification of molecules by breaking them up under controlled conditions so that they form characteristic fragments, which, taken together, provide a unique fingerprint of the original (parent) molecule. Unknown molecules can be identified by comparison with libraries of mass spectra. See Section 3.4.

ppb - parts per billion. Likewise **ppm** is parts per million, **ppt** parts per trillion and **ppq** parts per quadrillion. See Box 1 p3 for a full explanation.

Precision - a measure of the spread of results (obtained from repeat independent measurements on identical samples). Note that precision is not the same as accuracy - see above and Box 7 p35. See also repeatability and reproducibility.

Proficiency testing scheme - a system of inter-laboratory trials through which laboratories can judge their performance. Typically an independent body (the scheme organiser) sends participating laboratories standard samples for analysis. The laboratories submit their results to the organisers who issue a report from which each laboratory can identify only its own results but can compare these with the anonymised results of all other participants. Examples include Food Analysis Performance Assessment Scheme (FAPAS), QM and IMEP. See Section 2.9.

Protocol - in a general sense this can be taken to mean a set of instructions for an analytical method or a procedure for particular laboratory operations. More specifically it can also mean a standard method which should be followed exactly when carrying out an analysis.

Qualitative analysis - determining whether a substance (the analyte) is present.

Quantitative analysis - determining the amount of a substance (the analyte) in a known amount of sample.

Recovery - proportion of the amount of analyte (which is present in or been added to the matrix) which is extracted and available for analysis. Most extraction procedures, particularly with a complex matrix like food, will leave some analyte behind, so that recovery is less than 100%. Incomplete recovery of the analyte from the sample will result in a bias in the results (i.e. the results will be consistently lower than the true value).

Reference material - a material or substance, one or more of whose property values (e.g. purity or amount of analyte present) are sufficiently well established to be used for calibration of an instrument or the assessment of a measurement method. A certified reference material (CRM) is a particular type of reference material. The difference between the two is that a CRM must be accompanied by a certificate stating the property values for the material and an estimate of the uncertainty associated with each property value.

Repeatability - a measure of the degree of agreement of results obtained from repeat analyses carried out on identical samples under the *same* conditions (e.g. same analyst, equipment and laboratory). Repeatability is not the same as reproducibility (see below).

Reproducibility - a measure of the degree of agreement of results obtained from repeat analyses carried out on identical samples under *different* conditions (e.g. different analyst, equipment and laboratory or after a long interval of time). Reproducibility is not the same as repeatability (see above).

Ring trial - an inter-laboratory trial of a method. Typically each participating laboratory will analyse identical samples using the same protocol. The purpose of a ring trial is to assess the method, and particularly its reproducibility, and not to assess the laboratory (c.f. Proficiency testing scheme which is designed to assess the performance of the laboratory). See Section 2.4.

Robustness - see ruggedness, which is the preferred term.

Ruggedness - the extent to which the results of an analytical procedure are affected by changes in the procedure. The more rugged a method is, the less sensitive it is to changes in the conditions of its use. To quote an extreme example, over-the-counter tests sold for use in the home have to be extremely rugged so that they perform well even when used by a wide range of people under very different circumstances.

Sensitivity - strictly speaking, sensitivity refers to the change in response (signal) in relation to the amount of analyte present: the more sensitive the method the bigger the change. Although the term sensitivity has also been used to refer to the ability of a method to detect an analyte (i.e. the more sensitive the method the lower the limit of detection), this use is discouraged by analytical chemists.

Specificity - the extent to which the method responds to (detects) the analyte being sought and 'ignores' other, perhaps closely related, analytes.

Standard solution - a solution of a pure substance of known concentration, often used to calibrate an instrument or provide a positive control as a check that the method is working.

Glossary

Standard curve - a graph relating the signal from an instrument to the amount of analyte, constructed using a series of standard solutions of known concentration. The amount of analyte in an unknown sample can then be determined by using the graph to convert the signal obtained for the sample to the amount of analyte present (see Section 2.6)

Survey - a progamme of analysis of specific foods (or other materials) for specific analytes, to determine the occurrence and levels of the analyte in the foods in question. For example, in the UK the Food Standards Agency commissions surveys of particular foods for contaminants, authenticity and composition as part of its monitoring, enforcement and consumer protection role (see Section 4.5).

Traceability - this term is used in different ways by different groups of people and the distinction is important in avoiding confusion:

- To an analyst, this term describes a property of the result of a measurement or the value of a standard, whereby it can be related to stated references (usually national or international standards) through an unbroken chain of comparisons, all having stated uncertainties. This is described further and illustrated in Box 10 (p48)
- Within a laboratory, the term traceability is sometimes used more colloquially to refer to the route along which samples pass from the time they are taken to the point at which they are analysed. This is best referred to as a 'chain of custody' to avoid confusion with the above accepted use of the term traceability.
- Within the food supply chain, the term traceability is widely used to refer to the documented path taken by specific raw materials and ingredients in the processing, manufacturing and supply of particular food products. For example, a manufacturer may wish to avoid the use of ingredients from genetically modified crops and to achieve this will source the relevant ingredients from suppliers who are able to provide documented evidence of the provenance of the ingredients they supply.

Validation - confirmation that a method can yield acceptable results when it is used in the way intended. Validation will include the use of reference materials and standards. See Section 2.4.

8. SELECTED REFERENCES, FURTHER READING AND WEBSITES

This section lists some of the references and other information sources from which the information presented here is drawn. The titles in bold are likely to be particularly useful as a next step in following up the subject with further reading.

AOAC International - Association of Official Analytical Chemists - see *www.aoac.org*

Baigrie, B. (Ed.) (2003) Taints and off-flavours in food. Woodhead Publishing. ISBN 1-85573-449-4

Barwick, V. and Prichard, E. (2003) Introducing measurement uncertainty. LGC. ISBN 0-948926-19-8

Bender, D.A. and Bender, A.E. (1999) Benders' dictionary of nutrition and food technology. Seventh edition. Woodhead Publishing. ISBN 1-85573-475-3

BS EN ISO/IEC 17025:2000 General requirements for the competence of testing and calibration laboratories. See *www.bsi-global.com*

BSI (British Standards Institution) - see *www.bsi-global.com*

CCFRA (1999) Food authenticity assurance: an introductory guide for the food industry. CCFRA Guideline No. 23. ISBN 0-905942-21-3

CCFRA (2002) Guidelines for the preservation of official samples for analysis. CCFRA Guideline No. 35. ISBN 0-905942-47-7

CCFRA (2004) UK Food Law Notes - a reference manual on UK food law. CCFRA.

Chaovanalikit, A. and Wrolstad, R.E. (2004) Total anthocyanins and total phenolics of fresh and processed cherries and their antioxidant properties. Journal of Food Science **69** (19) FCT67-FCT72.

Codex (1999) Recommended methods of analysis and sampling. Codex Standard 234-1999. Available from *www.codexalimentarius.net*

Crosby, N.T. (1996) Sampling and sample plans for food surveillance exercises. pp1-31 in Progress in Food Contamination Analysis (Edited by J. Gilbert)

References

D'Mello, J.P.F. (2003) Food safety: contaminants and toxins. CABI Publishing. ISBN 0-85199-607-8

Downward, K. (2004) Mass spectrometry: a foundation course. Royal Society of Chemistry. ISBN 0-85404-609-7

Edwards, M.C. and Redpath, S.A. (1995) Guidelines for the identification of foreign bodies reported from food. CCFRA Guideline No. 5.

FAPAS® - Food Analysis Performance Assessment Scheme - see *http://ptg.csl.gov.uk/schemes.cfm* Provides useful information on some of the schemes in question, their purpose and how they work.

Fayle, S.E. and Gerrard, J.A. (2002) The Maillard reaction. RSC food analysis monographs. Royal Society of Chemistry. ISBN 0-85404-581-3

Food Standards Agency (2002) McCance and Widdowson's The composition of foods. Sixth summary edition. Royal Society of Chemistry. ISBN 0-85404-428-0

Food Standards Agency (2004) Practical sampling guidance for food standards and feeding stuffs. Available from *www.food.gov.uk*

Food Standards Agency website: www.food.gov.uk A useful source of information on current and past food chemical contamination, composition and authenticity issues, including UK Government food surveillance programmes.

Hill, H.C. (1972) Introduction to mass spectrometry. Second edition. Heyden. ISBN 0-85501-038-X

Hughes, E., Arnold, J., Bishell, K. and Farrar, E. (2004) If we get it wrong, everyone else is wasting their time. VAM Bulletin, Issue 31, pp7-8.

Hutton, T. (2001) Food manufacturing: an overview. Key Topics in Food Science and Technology No. 3. CCFRA. ISBN 0-905942-35-3

Hutton, T. (2002) Food chemical composition: dietary significance for food manufacturing. Key Topics in Food Science and Technology No. 6. CCFRA. ISBN 0-905942-50-7

Hutton, T (2003) Food packaging: an introduction. Key Topics in Food Science and Technology No. 7. CCFRA. ISBN 0-905942-61-2

Hutton, T (2004) Food preservation: an introduction. Key Topics in Food Science and Technology No. 9. CCFRA. ISBN 0-905942-69-8

IMEP - International Measurement Evaluation Programme - see *www.irmm.jrc.be*

Jewell, K. (2001) Designing and improving acceptance sampling plans - a tool. CCFRA Review No. 27. CCFRA.

Jones, L. (2000) Molecular methods in food analysis: principles and examples. Key Topics in Food Science and Technology No. 1. CCFRA. ISBN 0-905942-28-0

Kirk, R.S. and Sawyer, R. (1991) Pearson's composition and analysis of foods. Ninth edition. Longman. ISBN 0-582-40910-1

Kristott, J. (2004) Reducing trans fats in confectionery. Oils & Fats International. March 2004, pp20-22.

Leadley, C., Williams, A. and Jones, L. (2003) New technologies in food preservation: an introduction. Key Topics in Food Science and Technology No. 8. CCFRA. ISBN 0-905942-63-9

Lees, M. (Ed.) (2003) Food authenticity and traceability. Woodhead Publishing. ISBN 1-85573-526-1

LGC/VAM (2003a) Meeting the traceability requirements of ISO17025: an analyst's guide. Second edition. LGC. ISBN 0-948926-20-1

LGC/VAM (2003b) In-house method validation: a guide for chemical laboratories. LGC. ISBN 0-948926-18-X

McLean, B. (1999) Meat and meat products: the calculation of meat content, added water and connective tissue from analytical data. CCFRA Guideline No. 22 ISBN 0-905942-17-5

McLean, B. and Drake, P. (2002) Review of methods for the determination of fat and oil in foodstuffs. CCFRA Review No. 37.

Mueller-Harvey, I. and Baker, R.M. (2002) Chemical analysis in the laboratory: a basic guide. Royal Society of Chemistry. ISBN 0-85404-646-1

Prichard, E. (2004) Introduction to measurement terminology. LGC. ISBN 0-948926-21-X

QM *www.qualitymanagement.co.uk* - background information on the QM range of proficiency testing schemes

Reilly, C. (2002) Metal contamination of food: its significance for food quality and human health. Blackwell Publishing. ISBN 0-632-05927-3

Salmon, S.E., Voysey, P.A., Williams, J.S. and Brown, H.M. (2003) Microbially generated spoilage odours during grain storage - factors affecting their formation. CCFRA Review No. 35. CCFRA.

References

Schutz, H.W. (1984) How to design a statistical sampling plan. Food Technology, September 1984, 47-50.

Smith, R. and James, G.V. (1981) The sampling of bulk materials. Royal Society of Chemistry. ISBN 0-85186-810-X

Stanley, R.P. and Knight, C. (2001) Prioritisation of pesticide residue analysis: a practical guide. CCFRA Guideline No. 32. ISBN 0-905942-37-X

Stockley, C. and Lloyd-Davies, S. (2001) Analytical specifications for the export of Australian wines. The Australian Wine Research Institute. ISBN 0-9577870-2-2

UKAS - the United Kingdom Accreditation Service website *www.ukas.com* Contains background information on various aspects of accrediation as well as specific details of UKAS schemes.

VAM - Valid Analytical Measurement website. *www.vam.org.uk* Contains much valuable information and advice on best practice in chemical analysis across all industry sectors, as well as details of courses and seminars, technical reports and best practice guides.

VAM Bulletin - contains articles on wide ranging aspects of chemical analysis, including food analysis, to illustrate the benefits of and promote approaches to best practice. Available from *www.vam.org.uk*

VAM (1996) The manager's guide to VAM. See *www.vam.org.uk* A useful introduction to and explanation of the principles of VAM.

Voysey, P. (1999) Guidelines for the measurement of water activity and ERH in foods. CCFRA Guideline No. 25. ISBN 0-905942-23-X

Watson, D.H. (2001) Food chemical safety: contaminants. Woodhead Publishing. ISBN 1-85573-462-1

Watson, D.H. (2004) Pesticide, veterinary and other residues in food. Woodhead Publishing. ISBN 1-85573-734-5

Wilson, K and Walker, J. (Eds.) (2000) Principles and techniques of practical biochemistry. Fifth edition. Cambridge University Press. ISBN 0-521-65873-X.

Wood, R., Foster, L., Damant, A. and Key, P. (2004) Analytical methods for food additives. Woodhead Publishing. ISBN 1-85573-722-1

Wood, R., Nilsson, A. and Walin, H. (1998) Quality in the food analysis laboratory. RSC food analysis monographs. Royal Society of Chemistry. ISBN 0-85404-566-X

ABOUT CCFRA

The Campden & Chorleywood Food Research Association (CCFRA) is the largest membership-based food and drink research centre in the world. It provides wide-ranging scientific, technical and information services to companies right across the food production chain - from growers and producers, through processors and manufacturers to retailers and caterers. In addition to its 1600 members (drawn from over 60 different countries), CCFRA serves non-member companies, industrial consortia, UK government departments, levy boards and the European Union.

The services provided range from field trials and evaluation of raw materials through product and process development to consumer and market research. There is significant emphasis on food safety (e.g. through HACCP, hygiene and prevention of contamination), food analysis (chemical, microbiological and sensory), factory and laboratory auditing, training, publishing and information provision. To find out more, visit the CCFRA website at *www.campden.co.uk*

ABOUT VAM

Analytical measurement is of major economic significance and essential to ensuring quality of life with over £7 billion spent annually on chemical analysis in the UK. However, evidence suggests that many analytical measurements are not fit for purpose and that poor quality data represent a major cost and risk to business and society.

The VAM Programme helps organisations in the UK to carry out analytical measurements competently and accurately. The Programme enables the UK to demonstrate the comparability of analytical measurements with those of its trading partners and provides working laboratories with the 'tools' needed to implement best practice and demonstrate the reliability and integrity of their results. To find out more visit the VAM website at *www.vam.org.uk*